LEARNING RESOURCES CTR/NEW ENGLAND TECH.
GEN   TT267.L54 1988
Lieberman, E   Modern soldering & brazing

3 0147 0001 3278 0

TT267   S0-BSS-488

Lieberman, Eli.

Modern soldering & brazing
 techniques

DATE DUE

OCT 0 8 1991			
MAY 3 1992			

DEMCO 38-297

MODERN SOLDERING & BRAZING TECHNIQUES

By Eli Lieberman

 **BNP** Business News Publishing Company
Troy, Michigan

888 -170

Copyright © 1988
Eli Lieberman

All rights reserved. No part of this book may be reproduced or transmitted in any form or by any means—electronic or mechanical, including photocopying, recording or by any information storage and retrieval system—without written permission from the publisher, Business News Publishing Company.

Administrative and Text Editor: Phillip R. Roman

Printed in United States of America
7 6 5 4 3 2 1

Library of Congress Cataloging in Publication Data
Lieberman, Eli.
    Modern soldering & brazing techniques.

    Includes index.
      1. Solder and soldering. 2. Brazing. I. Title.
II. Title: Modern soldering and brazing techniques.
TT267.L54   1988   673'.356   87-30002
ISBN 0-912524-43-X

CONTENTS

Acknowledgements

Thanks to my son Ben who offered so much assistance in the preparation of this work, and to Ms. Sylvia Duerksen who drew the illustrations in the soldering section of this book.

I also want to acknowledge the assistance of all the manufacturers who contributed the material which made this book possible: Engelhard Corp.; Goss, Inc.; Handy and Harman; Imperial Eastman Div. of Clevite Industries; J. W. Harris Co.; L-TEC Welding & Cutting Systems; Mueller Brass Co.; NIBCO, Inc.; Ritchie Eng. Co. Inc.; Rigid Tool Co.; T-Drill Co.; Uniweld Products, Inc.; Westinghouse and Wingaersheek Div. of Victor Equipment Co. Special thanks also to the Copper Development Association and the American Welding Society.

The photograph depicted on the cover is provided through the special courtesy of the Engelhard Corporation.

—E.L.

DISCLAIMER

This book is only considered to be a general guide. The author and publisher have neither liability nor can they be responsible to any person or entity for any misunderstanding, misuse or misapplication that would cause loss or damage of any kind, including material or personal injury, or alleged to be caused directly or indirectly by the information contained in this book.

PREFACE

What constitutes art? Typically, the reaction is that it is ir-reproduceable—one of a kind. Art, by definition, is the ac-quisition of a skill by observation, study or experience—yet art commonly implies an unanalyzable, creative power whose results can rarely be reproduced. However, since the course of this text has endeavored to analyze the processes of soldering and brazing, these metal joining techniques might more properly be considered something other than an art.

By its nature, the term craft implies skill in planning and effectively executing a task (i.e., workmanship) to con-sistently produce and reproduce quality results time and again. The operative phrase in this assertion—*consistently produce and reproduce*. By virtue of the fact that soldering and brazing (especially in the refrigeration and air condition-ing trades) must consistently yield resilient, tightly sealed connections to be effective, it is logical to consider soldering and brazing as crafts rather than arts.

Thorough knowledge of the soldering and brazing proc-esses requires a blend of both science and hands-on experi-ence. Science explains the nature of a soldered or brazed joint and how it forms. It explains how capillary action works and why alloys behave as they do. And it also reveals why a reducing flame should be used rather than an oxidizing flame. In short, science explains how and why soldering and brazing works.

However, experience is necessary in applying the science. Skill must be developed. Knowing how and precisely when to feed filler metal into a clearance space to take full advantage of capillary action is essential for producing satisfactory connections. And certain skill must be acquired when working with brazing alloys in the *pasty* range to affect satisfactory results.

This book has brought together theory and practical how-to information as it pertains to the specialized applications of soldering and brazing in the refrigeration and air conditioning trades. Consequently, the information presented between these covers should aid newcomers to the refrigeration and air conditioning service industry to rapidly acquire proper soldering and brazing skills. To foster improved methodologies of the more experienced service personnel, the scientific aspects have been included herein to help shed some light on why certain metal joining techniques are used and recommended over others and just how these effect the craft of soldering and brazing.

Eli Lieberman

PART I

SOLDERING

INTRODUCTION

The general subject of metal joining is so vast that it is difficult to merely keep abreast of the current literature being produced on the topic. Fortunately for those employed in the refrigeration and air conditioning trade, only a narrowed scope of soldering—dealing primarily with joining copper tube and, in more recent times, field repair of aluminum components, must be addressed for immediate applications. Even though the scope of interest on the topics of soldering and brazing has been narrowed to a great extent for the purposes of this text, joining copper tube for refrigeration and air conditioning applications requires special precautions and techniques that are not necessarily applicable to other trades. Conditions of pressure, temperature and vibration are more severe than those encountered in normal plumbing installations for example. In addition, the requirement for a contamination-free hermetic system can add many complications to the installer's job.

The primary objective of this book is to bring together, in an easily comprehendible form, modern soldering and brazing information and techniques as it specifically pertains to joining copper tubing. Although directed primarily towards the concern of refrigeration and air conditioning service personnel, this book should also prove to be of value to any field operative who must tackle the task of joining copper tubing.

From a historical aspect, as the refrigeration trade developed in the thirties and forties around the use of copper

1

tube and pipe, it was only natural to employ the terminology and techniques of the plumbing and copper metal trades. Originally, lead had been referred to as a *soft solder* and tin as a *hard solder*. As soldering and brazing filler metals began to proliferate, a trend developed to designate tin-lead solders as *soft solders* and brazing filler metals as *hard solders*.

To make matters confusing, many service personnel — oblivious to sound metallugrical and technical rationale, mistakenly refer to the moldable wire form of tin-lead solder as *soft solder* and to rigid silver-content brazing rods as *hard solder* merely due to the flexability factor of the substances.

However, most hard solders contain silver, so the term *silver soldering* also came into use and eventually became synonymous with hard soldering.

The situation changed in the early fifties with the introduction of silver-bearing soft solder which rapidly became an important soft solder in the refrigeration trade. The use of the term *silver soldering* in referring to brazing then became obsolete.

Once a specific terminology takes root in the trade literature and industry jargon, change becomes very difficult. Nevertheless, for those now entering the trade, and for all future publication applications, an honest attempt should be made to use more current and accurate terminology in order to belay the confusion that has resulted from the application of misnomered terminology.

Terminology

Copper tube is joined by two basic methods:
- soldering
- brazing

In order to eliminate some of the confusion which exists, suggested definitions and descriptions of these terms should be examined briefly.

- **Soldering**

In soldering, an overlapped joint of copper tubing is bonded by a low melting alloy referred to as solder. The solder melts at a much lower temperature than the copper base metal and is consequently drawn into the joint by capillary action. The strength in a soldered joint depends upon the alloying action between the solder and overlapped copper sections. Spacing is critical since solder alone has a very low tensile strength.

The most important distinguishing aspect in the process of soldering is that the metal joining takes place at a temperature below that which affects the properties of the copper (base metal). Since its physical properties are not altered by the heating, there is no color change in the copper (i.e., no black copper oxide scale is formed). In addition, no significant annealing of work hardened copper tube takes place. For convenience, soldering has been defined by the American Welding Society (AWS) as taking place below 840°F. But as a practical matter, considering the solders used in the refrigeration trade a figure of 600°F is more accurate.

The term "soldering" is used here in its correct metallurgical sense. The original meaning was intended to describe a metal joining operation employing a bonding alloy which melted at a temperature low enough so as not to alter the properties of the base metal. It should be noted that the terms *soft* and *hard* have purposely been omitted in the course of this text to eliminate unnecessary and often confusing verbiage.

- **Solder Alloys**

Solders currently used in the refrigeration and air conditioning trade are all simple alloys of tin and another metal. The three basic alloys are: tin-lead, tin-antimony and tin-silver. During any discussion, solders should be designated by first naming the metals in the alloy, followed by the percentages of tin and the other metal in that order.

The most common solders used in the refrigeration

trade, including solders used for water and drain line applications, are listed below.

tin-lead	40/60
tin-lead	50/50
tin-lead	60/40
tin-antimony	95/5
tin-silver	94/6
tin-silver	96/4

When a solder is sold under a particular brand name, the solder can be so specified. For example, tin-silver solders are sold by the Harris Company under the trade name of STAY-BRITE. The Engelhard Corporation markets tin-silver solders under the trade name SILVABRITE.

- **Brazing**
   Copper tube may be joined in a brazing process using a brazing filler metal which melts above 840°F but below the melting point of copper. Brazing filler metals used in the refrigeration trade generally melt at about 1,300°F. The same fittings are utilized in both soldering and brazing because, in both instances, capillary action is relied upon to draw molten filler metal into the space between the fitting and tube. One differentiation exists between the two processes in that the higher temperature required by brazing affects the physical and chemical properties of the copper where, as previously indicated, soldering does not.

   In brazing, the copper tube is annealed (a black copper oxide residue is formed) and the copper may be weakened by increased grain size. To minimize these detrimental effects, a brazing operation should be accomplished as rapidly as possible.

   Even though the base copper is weakened to a slight degree, a brazed joint is ultimately stronger than a soldered joint because the tensile strength of the brazing filler metal is much greater than solder. In addition, joint clearance is not as critical in brazing as it is in soldering since the strength of the brazing filler metal can be relied upon to bridge greater clearances than those allowed in a soldered joint.

- **Brazing Filler Metals**
  Brazing filler metals used to join copper tubing are all alloys of copper and one or more other elements. The principle copper alloys employed are:
  copper-silver
  copper-phosphorus
  copper-silver-phosphorus
  copper-silver-phosphorus-tin
  copper-silver-cadmium-zinc
  It has become customary to specify a brazing filler metal by its brand name rather than by its metal composition. In view of the relatively large number of brazing filler metals available on the market and the wide variety of alloys in use, it is best to continue the present practice of referring to the filler metals by brand names.

- **Sweat**
  The term "sweat" is used in a general sense to describe a joint which may be made by soldering or brazing. For the most part it is employed to draw a distinction between a mechanical type of connection and one made by soldering or brazing. For example, a catalog may list a shut-off valve as having sweat connections to distinguish it from one having flare or pipe connections. When sweat is specified, the instructions should clearly indicate whether soldering, brazing or either one can be used. In this regard many refrigeration valve catalogs inappropriately use the terms soldering and brazing interchangeably. Certain types of valves easily sustain heat damage and, as a result, require special handling procedures when being brazed. Consequently, it is important to determine exactly which metal joining process (soldering or brazing) the manufacturer recommends for a particular connection when the term "sweat" is referred to in the product information.

CAPILLARY ACTION

The refrigeration and air conditioning trade is fortunate in that it employs easily installed and repaired piping systems by design. The high pressures and temperatures which exist in refrigeration piping could well have required cutting and threading pipe, using flanged connections or even welding. Part of the credit goes to the availability of soft copper tubing in rolls and hard drawn copper tube in straight lengths. The "red metal" has served the trades well for many years. It is reliable, easily installed, and may be joined in a permanent leak-proof bond.

A good share of the convenience of tubing use is due to the naturally occurring phenomenon of capillary action. It is capillary action that makes soldering and brazing possible by enabling molten solder or brazing filler metal to be drawn into a joint making a permanent bond. Without capillary action, solder would have to be poured into a joint similar to the way in which lead is poured into the bell of a cast iron pipe to form a leakproof seal.

Capillary action can be explained by referring to a tried and reliable high school science demonstration. A number of small bore capillary tubes are arranged vertically in a vessel filled with water (Figure 1-1). As the diameter of the tubes decreases, the water can be observed to rise higher above the surface of the water in the vessel.

The water rises in the tube because it wets the glass and adheres to it. A concave depression called a *meniscus* is created on the top surface of the water because the water adjacent to the glass is attracted more strongly than the water located in the central region. The adhesion and surface tension combine to "pull" the water column upward.

Capillary action is not only essential to soldering but also plays an important role in nature. It enables sap to rise in trees, blood to flow to the cells in the body, and moisture to move through the soil.

What has been said concerning capillary action of water also applies to capillary action of molten solder. Figure 1-2

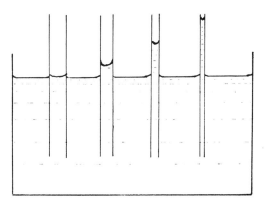

Figure 1-1. Capillary action of water in small diameter tubes.

shows a typical solder joint having a tube-end portion inserted within an expanded socket. A clearance space exists between the walls in the coupling. Cap-off the bottom end of the tube and insert the joint in a vessel of water as demonstrated in Figure 1-1. The water rises and fills the clearance space between the walls.

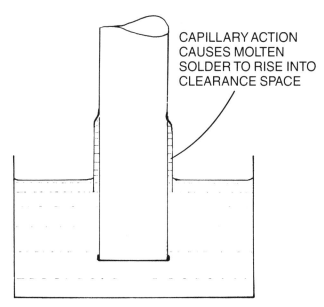

CAPILLARY ACTION
CAUSES MOLTEN
SOLDER TO RISE INTO
CLEARANCE SPACE

Figure 1-2. Capillary action in solder fitting.

Now suppose the liquid in the vessel of Figure 1-2 is molten solder. A similar action takes place. Solder is drawn up into the clearance space by capillary action. In fact, this type of molten solder bath is used in conjunction with a conveyor system to mass produce soldered joints in factory settings.

Solder does not have to rise up out of a solder bath but can be applied out in the open at any point in the joint opening and in any position.

SOLDER CLEARANCE SPACE

At this point a clear distinction must be drawn between soldering and brazing. Although the two metal joining processes have many features in common, some differences are so significant that they must be treated separately. What follows relates mainly to soldering. Common features of soldering and brazing, such as cleaning copper tubes and fittings and tube-end preparation, are apparent. Brazing proper will be addressed in a part two of this book.

Figure 1-3 shows a schematic cross sectional view of a typical soldered joint connecting two copper tubes.

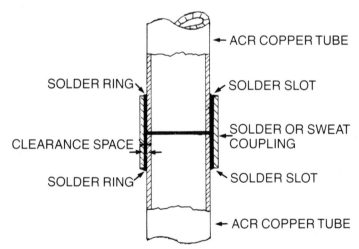

Figure 1-3. Typical solder coupling.

The nomenclature on the drawing is defined as follows:

- **ACR copper tube** — This copper tube is produced specifically for the air conditioning and refrigeration industry, hence its abbreviation ACR tube. It is available in soft annealed coils or hard-drawn lengths. Processed with special attention to cleanliness and dryness, ACR tube is intended for use in the field with special fittings for connections, repairs and alterations in air conditioning and refrigeration installations.
- **Solder or sweat coupling** — This coupling is a straight solder fitting dimensioned to receive a pair of ACR tubes with proper clearance to form a soldered joint.
- **Clearance space** — The distance between the outside surface of the tube and the inside surface of the solder fitting constitutes the clearance space, and it is expressed in thousandths of an inch. Clearance space is also a factor used in determining the clearance volume which is equal to the total volume between the fitting and tube that must be filled with solder.
- **Solder slot (mouth)** — The slot is an annular opening between the tube and coupling-end into which molten solder is fed.
- **Solder ring** — An annular ring of solder around the solder slot which appears after the clearance volume has been completely filled to overflowing with molten solder.

The tube-ends are inserted in a solder coupling manufactured to tolerances set by ANSI B16.22 of The National Standards Institute. The reason for setting rigid specifications in the manufacture of solder fittings is to ensure a correct clearance between the tube and fittings. This clearance must fall within the range of .002- to .006-inch in order to develop the optimum capillary action required to draw the molten solder into the clearance space of the fitting.

If the clearance is under .002-inch, the molten solder may have difficulty in effectively penetrating the clearance space. If the clearance is greater than .006-inch, the solder may not completely fill the clearance space and a weak joint may result.

When the clearance is correct, the force of capillary action is so much greater than the force of gravity that a soldered joint can be made in any position relative to the gravitational field. Assuming the soldered joint in Figure 1-3 is made in the vertical position, the solder flows up the bottom connection with the same ease that it flows down the top one.

Another reason for maintaining a tight clearance is to provide a joint having good mechanical strength. Solder itself is weak in tensile and shear strength; the joint must rely on the base metal for its strength. Solder serves the same purpose as does adhesive or glue when joining two pieces of wood. The adhesive itself is quite weak but, if spread in a thin layer, the wood fibers joined by the adhesive supply the necessary strength to hold the joint together once the adhesive sets and cures. In a similar fashion, if the solder layer is thin enough, some of the strength of the copper base metal can be imparted to the solder by an alloying action between the copper and solder.

TUBE CUTTING AND TUBE-END PREPARATION

In order to achieve a good fit with standard sweat fittings. a tube-end must be cut squarely—without inducing any out-of-roundness. The preferred method of cutting copper tubing is with a tubing cutter. Small portable tubing cutters are available which can handle tube sizes that range from ⅛-inch OD to 4 ⅛-inches OD.

Although called a tubing cutter, the metal is not actually removed in the same sense that the teeth of hacksaw blade saw away at metal by removing minute metallic chips. Instead, with a tubing cutter, the blade of a cutting wheel is forced against the external surface of the copper tube and rotated around the circumference of the tube. By rotating the tool, metal is displaced to either side of the blade's edge until the tube parts.

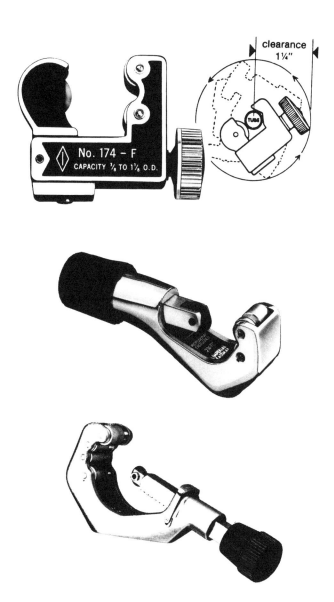

Figure 1-4. Tubing cutters of various sizes (Courtesy, Imperial Eastman).

There are many advantages in using a tube cutter over other methods (such as using a hacksaw).

- The tube-end is always cut squarely because of the centering action of the rollers and cutting wheel.
- No copper chips result (which are apt to enter the system and damage equipment).
- A tubing cutter can be operated in any position without clamping down the tubing.
- It is a durable tool that is conveniently carried in a tool box.

The only obvious disadvantage associated with the use of a tube cutter on copper is that it may produce a burr or ridge which must be removed in a separate operation.

On soft copper tubing a burr is often formed on the side lip of the tube and must be removed because it tends to add resistance to the flow of refrigerant. On hard drawn copper tubing a ridge is raised on the outside and, when compared to soft copper tubing, a somewhat smaller burr is formed on the inside lip. The outside ridge that results on hard drawn tubing must also be removed because it may interfere with seating the tube-end in the fitting socket. Figure 1-5 shows an enlarged view of the burrs and ridges formed on the tube-end as a result of the tubing cutter action.

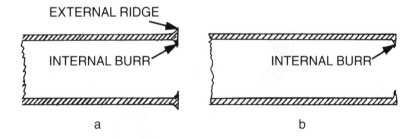

Figure 1-5. a) Ridge and burr on hard drawn copper tubing; b) A burr on soft copper tubing.

The size of the burr and the ridge can be minimized by maintaining a sharp cutting wheel. A dull wheel requires more effort on the part of the operator, while at the same time displacing more metal and increasing the size of the burr and ridge. Frequently inspect and change the cutting wheel when dull. Also, periodically inspect the cutting wheel shaft screw for wear. A worn cutting wheel shaft screw allows the cutting wheel to wobble and typically generates a spiral thread screw on tubing instead of producing a closed, concentric cut that yields a square tube-end.

Burr and ridge formation can also be reduced by not overdriving (overtightening) the cutting wheel. After a first initial heavy cut, the cutter should be rotated while advancing the handle no more than a ¼ turn per rotation. It is better to rotate the cutter a few extra turns to get a smooth cut rather than overdriving the cutter which results in a heavy burr build-up.

A number of tools are available to remove the burrs and ridges produced by the use of a tubing cutter. Figure 1-6 shows an inner-outer reamer which is suited to handle tube sizes ranging from ⅛-inch to 1 ⅜-inches. An inside conical reamer on one side of the tool removes internal burrs while an outside conical reamer on the other side of the tool removes the outside ridges. This tool has an advantage in that it can effectively remove inside and outside burrs through a large range of tubing sizes. It is also a durable, compact and effective tool.

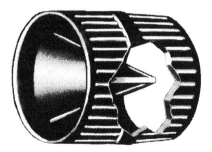

Figure 1-6. An inner-outer tubing reamer (Courtesy, Imperial Eastman).

Heavy duty tubing cutters are usually equipped with a flat triangular reamer which is attached to the side of the tool and folds out and locks open for use. Rotating the reamer in the tube-end a number of times gradually shaves away the internal burr. In addition, a number of other tools and techniques may be employed for burr and ridge removal. Heavy duty cutting knives are available which can be used to actually cut the burr out of the tube-end. When the knife blade is inserted into the tube-end and rotated in a circular cutting motion, the burr is removed as a copper sliver. A rat-tail file may also be used to remove the inside burr while a flat file is effective in removing the outside ridge. Remember, all copper filings should be removed from the tube-end to prevent their entering the system.

In an installation where many connections must be made, a tube reamer (as illustrated in Figure 1-7) driven by a battery powered electric drill may prove to be a significant time saver.

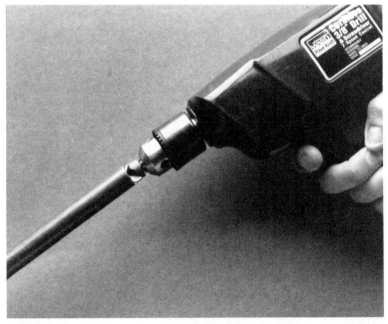

Figure 1-7. Tube reamer driven by battery powered electric drill.

A hacksaw with a 32-teeth/inch blade can also be used to cut tubing. Although the major disadvantage in using a hacksaw is that it is difficult to achieve a square cut without the use of a special tube sawing vise. The copper filings which result from the sawing operation also tend to work up inside the tubing. If the filings accumulate, they can plug strainers and contaminate the oil in a compressor.

Out-of-Roundness

Copper tubing is manufactured to tightly controlled tolerances at the factory but can easily be distorted during handling, packaging and shipping. The most frequently encountered distortion is out-of-roundness. This condition occurs most often in ¾-inch OD and ⅞-inch OD annealed tubing packaged in rolls. The coiling process itself introduces a certain degree of out-of-roundness which is exaggerated further when the tubing is unrolled and strung in place during installation. With hard drawn tubing, out-of-roundness results from rough handling—particularly at the tube-ends, caused by dropping on a concrete floor.

Out-of-roundness is most evident when a serviceman experiences difficulty in inserting the tube-end into a fitting socket. Concentricity (roundness) can be restored by inserting the cylindrical guide of a swaging tool into the tube-end and gently driving it into the tubing until the lip of the tube is seated on the tool's flaring section (Figure 1-8).

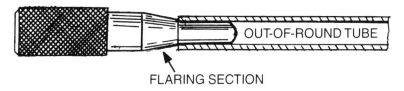

OUT-OF-ROUND TUBE

FLARING SECTION

Figure 1-8. Swaging tool used to restore concentricity to tube-end.

Sizing tools are available for restoring more severely deformed tube-ends to their original condition (Figure 1-9).

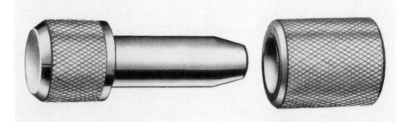

Figure 1-9. Sizing tool used for restoring concentricity to damaged tube-ends (Courtesy, NIBCO).

The problem presented by the use of sizing tools is that a separate tool is required for each tubing type and size. Rather than carry a set of sizing tools, many servicemen prefer to improvise by restoring tubing with swaging tools. However, there is merit in carrying a set of three sizing tools for the popular 5⁄8-, 3⁄4- and 7⁄8-inch ACR tubing sizes.

Generally speaking, when an out-of-round tube-end is restored and inserted in the fitting socket, a satisfactory soldered joint can be made. However, under certain unusual conditions of distortion between the tube-end and fitting socket, a gap may appear in the clearance zone which is far in excess of the .006-inch clearance for optimum capillary action. In this sort of situation, the solder may show a tendency to run out of the wide gap. However with carefully controlled heat and solder applications, the gap can still be filled satisfactorily.

SURFACE PREPARATION

MECHANICAL CLEANING

With the matter of proper clearance having been determined by the use of standard solder fittings and proper tube end preparation, the first step in making a soldered connection is to thoroughly clean the joining surfaces. In order for solder to adhere to the base metal and for capillary action to take place, solder must *wet* the surface. Wetting is the permanent film left on a metal surface by a molten solder. Wetting is partially chemical and partially mechanical in nature. The wetting action depends on the nature of the base metal and the solder. Copper is easily wetted by solders containing tin, while aluminum is wetted with more difficulty. No intervening layers of foreign material can exist between the base metal and the solder. Layers of dirt and grease must be removed along with any metallic oxides that may have formed prior to soldering if proper wetting is to occur.

Abrasive cloth is excellent for cleaning the outside surface of tubing. The cloth is generally available in 25-yard rolls, 1 ½-inches wide, and is usually packaged in a convenient dispenser. In use, a strip of cloth is worked back and forth over the tubing with a buffing action, similar to that used when shining shoes, until a bright copper color is achieved (Figure 1-10).

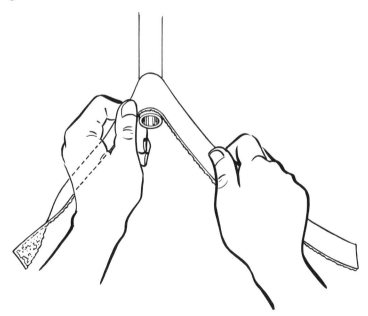

Figure 1-10. Abrasive cloth used for cleaning copper tube.

Generally, spiral wire brushes are recommended for cleaning the inside surfaces of most fittings. The only problem with using spiral brushes for this purpose is the large inventory required to handle the diversity of fitting sizes.

Abrasive cloth can also be used to clean internal fitting surfaces, but it should only be used as a back-up in the event the brush size required for a job is not readily available or is missing from a brush set. To facilitate cleaning of ¼-inch to ⅝-inch fittings, a piece of abrasive cloth can be rolled into a tight cylinder slightly less than the inside diameter of the fitting. When inserted into the fitting and released, the rolled up cloth expands against the inner wall. The inside surface can then be cleaned by simply rotating the abrasive cloth cylinder back and forth (Figure 1-11). On fittings larger than ⅝-inch, a piece of abrasive cloth under the thumb or wrapped around the forefinger can be worked around the inside surface to remove oxide layers as illustrated in Figure 1-12.

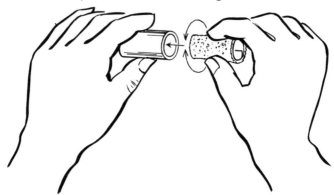

Figure 1-11. Cleaning a fitting with an abrasive cloth roll.

When using the abrasive cloth, the abrasive particles which result from the cleaning operation must be removed by wiping, brushing, tapping or blowing—depending upon the circumstances. If allowed to remain on the cleaned surfaces, the particles can interfere with the soldering operation. For obvious reasons, make certain that abrasive particles are not permitted to enter the refrigeration system.

Figure 1-12. Abrasive cloth used for cleaning copper fittings.

Abrasive cloth should not be used to clean a connection where the there is a chance of abrasive residue falling into the system before it can be safely removed. For example, abrasive cloth should not be used on the upward directed suction and discharge stub tubes of a compressor since the abrasive residue falling to the bottom of the stub tube bend can be difficult to remove. A spiral wire cleaning brush should be used in this sort of situation because it can be manipulated with a upward, rotary motion that picks up the residue and carries it out of the fitting.

Figure 1-13. Spiral wire brush used for cleaning fittings.

CHEMICAL CLEANING
— Flux

Even though mechanical cleaning may expose a shiny copper or brass colored base metal, an invisible oxide layer still exists that interferes with the wetting process. Consequently, this oxide layer must be removed chemically and prevented from reforming when the metal is heated. A soldering flux is employed for this purpose. Flux is a chemical that possesses the following properties:

- It is fluid at soldering temperatures and capable of adhering to the base metal.
- It is chemically active and capable of dissolving metal oxides.
- It is able to prevent oxides from reforming.
- It is capable of being displaced by the molten solder as it wets the base metal.

Zinc chloride, one of the few chemicals that possesses all of the above described properties, is the flux preferred when joining copper tubing. Since it melts at about 503°F, which is slightly above the melting point of many solders, it is usually mixed with ammonium chloride to lower its melting temperature. Zinc chloride flux comes in a paste and liquid form. Both types have advantages depending on the application.

To make a paste flux, zinc chloride and ammonium chloride are mixed in a quantity of petroleum jelly. In use the surfaces to be soldered are covered with a thin layer of paste. A toothbrush is convenient to use for this purpose on the larger size tubes and fittings (Figure 1-14).

On the smaller tube sizes, such as ¼-inch, the wire solder itself can be used as a tool to spread the paste around. Never spread paste with your finger. Dirt and oily skin secretions contaminate the metal surface and may interfere with the fluxing action.

When necessary, spread flux paste sparingly. Excess paste can easily run into a refrigerant system and become a

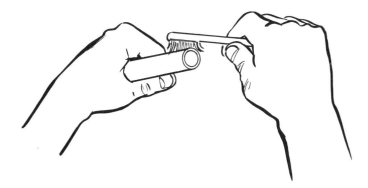

Figure 1-14. Using a soft toothbrush to coat tube surface with flux.

contaminant which is difficult to remove. An expansion valve is known to stick because of waxy deposits that occur around its orifice. This waxy build-up is typically attributed to excess soldering paste that's been allowed to enter the system due to sloppy soldering methods.

The same container of flux should not be used and carried in the service truck for a period of more than a year at most. During hot weather, for instance, the paste tends to melt. This melt down often results in an uneven distribution of the active chemicals. The balance can be restored by simply mixing the paste before reusing. Still, it's best to renew paste flux periodically to prevent any contamination problems. With extended use, foreign materials such as dirt and abrasive grit can accumulate in the container and may reduce the effectiveness of the flux.

Liquid flux contains the same active ingredients as paste flux and in many ways is more convenient to use in refrigeration work. Typically packaged in plastic squeeze bottles, liquid flux can be dispensed a drop at a time. This is an advantage because liquid flux can be applied to the joint surfaces in a more conscientious manner lessening the chance of its working into and contaminating the system.

An effective way to apply liquid flux is by means of cotton swabs similar to those sold in drugstores. Saturate the cotton

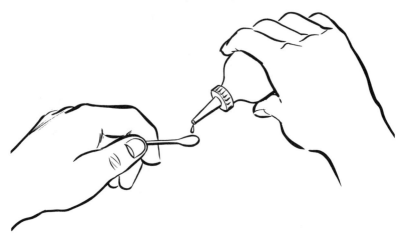

Figure 1-15. Saturating the swab with liquid flux.

with a few drops of flux (Figure 1-15), rather than squirting directly on the joint surfaces, and swab the inside of the fitting and the outside of the tube as show in Figure 1-16. This technique supplies the correct amount of flux and avoids the problem of using too much flux. The cotton swabs should be changed frequently because they wipe away particles which result from mechanical cleaning.

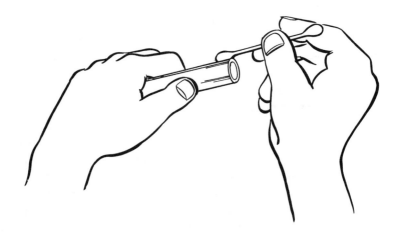

Figure 1-16. Swabbing the tube surface with liquid flux.

Another advantage of using liquid flux is that its dispenser keeps it free of contaminates. In the case of paste flux, brushes must be dipped into a flux container which contaminates the entire supply whereas liquid flux is dispensed only as necessary from the bottle. In addition, liquid flux does not exhibit the tendency to separate out during storage and use as does the paste form.

If necessary, paste flux can remain on joint surfaces for an hour or two before soldering because it is not as chemically active as its liquid counterpart. When using liquid flux, however, perform the soldering operation immediately following the application of the flux.

The role of flux in making a soldered joint is explained with reference to Figure 1-17. This figure is a schematic view of a soldered joint that shows a greatly exaggerated clearance space for the sake of discussion.

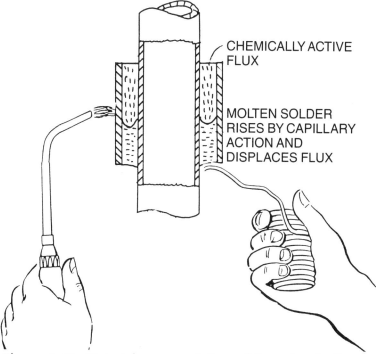

CHEMICALLY ACTIVE FLUX

MOLTEN SOLDER RISES BY CAPILLARY ACTION AND DISPLACES FLUX

Figure 1-17. Pictorial representation of the role of a flux in a soldered connection.

After the joint is cleaned, fluxed and assembled, it is heated to bring the flux to its boiling temperature. The heat acts as a catalyst to make the flux chemically active. Once active, the flux dissolves the thin film of metal oxide that remains after the mechanical cleaning operation. At the same time, the hot flux coats the chemically clean base metal to inhibit any reaction with air and the subsequent reforming of metal oxide. When the joint temperature reaches about 540°F , the wire solder touching the soldering slot melts and is drawn into the clearance space by capillary action. The molten solder wets and clings to the base metal forming a meniscus on the top surface. The surface tension of the meniscus "pulls" the molten solder column upward while simultaneously displacing the flux from the wall. The rising column of solder strips the flux from the wall and maintains an air-tight fluid seal to prevent oxide reformation. The process ends when the solder has displaced all the flux and the clearance volume is completely filled with solder.

Electrical Connection Fluxing

One precaution must be observed. Never use zinc chloride flux, either in paste or liquid form, when making electrical connections. Any excess zinc chloride flux will remain active and corrode fine electrical wires while degrading insulation.

For soldering electrical connections, use a specially prepared wire solder with a resin filled core. At room temperature it is inactive and does not cause any corrosion. When it is heated to soldering temperature, it becomes sufficiently active to serve as a flux.

SOLDER FEEDING AND HEATING

Applying Heat

After the preliminary cleaning and fluxing of metal surfaces, mechanically assemble the joint in preparation for soldering. In order to prevent oxides and other contaminants from reforming on the base metal, it is best to complete a soldering operation as quickly as possible.

Soldering is best performed with a low-heat flame. An air-acetylene, air-propane or other air-fuel gas torch is generally used. Oxy-acetylene is not typically recommended for soldering, but if it is the only heat source available on the jobsite it can be used provided an adjustment is made so a low-heat flame results.

The primary skill in soldering is in mastering the technique of correlating heating with solder feeding. Although an essential skill, it is one that can only be acquired through hands-on experience. Nevertheless, some guidelines should prove to be helpful in the acquisition of this skill.

There are two main aspects which need to be recognized when heating a joint in preparation for soldering:

- the size of the flame
- the location of the flame

Since soldering occurs at a temperature well below the flame temperature of all the fuel gases, factors relating to the properties of the fuel gas and the nature and shape of the flame are not of critical importance. These factors, however, are of considerable importance in brazing and will be covered in greater detail in the brazing portion of this text.

The soldering flame must be large enough to deliver sufficient heat (measured in British Thermal Units or Btu) to raise the temperature of the joint to the melting point of the solder. In soldering tubing up to ½-inch in diameter a light tip with a heating capacity of about 5,500 Btu is sufficient. Of course, this same tip does not generate enough heat to sweat a 1 ½-inch joint.

In the case of the larger 1 ½-inch joint, the copper adjoining the coupling tends to conduct the heat away from the soldering point faster than it can be delivered so correct soldering temperature cannot be attained. Instead, a medium tip with a capacity of about 11,000 Btu is required in that instance. Considerable overlap exists in the range of operation of the various tips, so tip selection is not critical. Considering that most copper tube soldering in the refrigeration trade will be below 2 ⅛-inches, two or three tips should easily handle the tube sizes commonly encountered.

For lack of a properly sized tip, an oversized tip should present no problem to personnel experienced in soldering. What is required is that the torch be played back and forth across the joint to quickly raise it to soldering temperature without causing any overheating. One way of describing this technique is to draw an analogy between a short-order cook and someone who rarely cooks. When rushed at breakfast time, a cook usually increases the stove's heat setting to its maximum and, using an extremely hot skillet, rapidly scrambles the eggs with a fork. Because the heat is at its maximum, the eggs cook in matter of seconds. Just as soon as the eggs are ready, the cook quickly tosses them out of the skillet. In contrast, someone who occasionally scrambles eggs for breakfast typically uses a much lower heat setting and slowly scrambles and tests the eggs until satisfied that they are done.

The situation is similar in soldering. To experienced personnel heat control poses no problem. A serviceman knows from experience which torch to use on a particular connection. If the flame turns out to be too large, experience has taught the serviceman how to play it around the connection to dissipate the excess heat.

For the beginner it is best to use a rated torch until the skill of heat control has been mastered. It is easier to compensate for slight underheating rather than for overheating.

Another important consideration is flame location. Initially a flame is directed back and forth across the solder joint until the flux becomes active. At that time the flame is aimed at a centralized location so that maximum heat is directed at the base of the joint. Solder is then fed into the solder slot to complete the operation. The reason a flame is held remote from the slot can be explained by referring to Figures 1-18 and 1-19.

Figure 1-18 shows a sweat coupling with the flame directed at a central location at the moment the solder begins to flow.

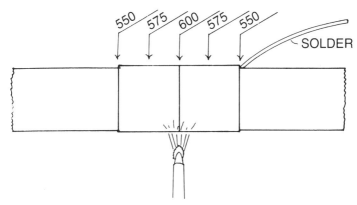

Figure 1-18. Soldering with flame at base of joint.

Note the surface temperature readings. The central part of the fitting where the heat is concentrated has reached 600°F with a falling temperature gradient down to the slot where it is only 550°F. A tin-silver solder is used which melts at 430°F. The solder will melt at the solder slot and be drawn into the joint where the temperature is even higher. The clearance volume is then completely filled with solder yielding a good solder bond.

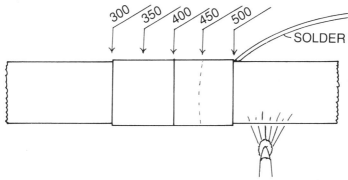

Figure 1-19. Soldering with flame away from base of joint.

On the other hand, Figure 1-19 shows the situation where the heat is directed away from the base of the joint at the moment the solder begins to melt.

In this case, the temperature gradient falls in the wrong direction and is too low at the bottom of the joint to properly draw-in the solder. As a result, the solder is only drawn into the joint part way (as shown by the dotted line) yielding an imperfect solder bond.

The optimum temperature for soldering is about 550°F. Starting with a cold joint, play the torch back and forth across the fitting and tubing to raise the temperature of the parts in a uniform manner (Figure 1-20).

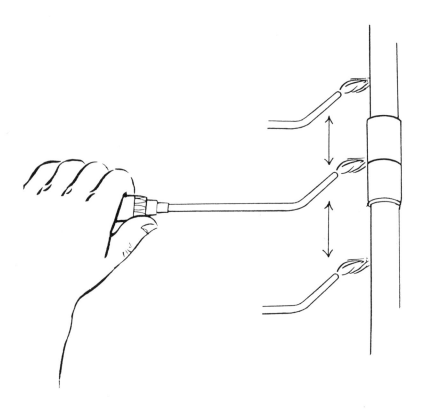

Figure 1-20. Raising the temperature of a joint in a uniform manner.

In a short time the flux will start to boil and become active. Once the flux begins to boil, concentrate the flame at a central portion of the fitting where the tube-ends join and rub the end of the wire solder against a small arc of the solder slot (Figure 1-21).

If it does not melt, the joint is not quite hot enough so you must repeat the rubbing procedure every few seconds. The solder should not be melted by the direct flame of the torch but rather should be melted by heat conducted through the base metals. As soon as the solder melts, keep feeding solder until a full solder ring forms around the solder slot.

Once the soldering operation has been completed, the joint should be allowed to cool thoroughly.

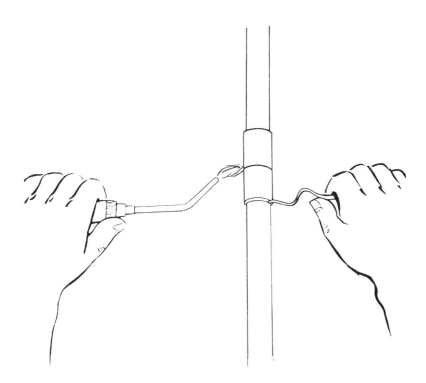

Figure 1-21. Testing for optimum soldering temperature.

Solder Ring

The ability to achieve and recognize a full solder ring is essential in mastering the craft of soldering. On a properly cleaned and fluxed joint, a full solder ring indicates that enough solder was drawn into the joint by capillary action to fill the clearance volume up to its brim. Lack of a visible solder ring may indicate a partially filled, weak joint.

If a joint is made in a horizontal position, a full solder ring may be recognized by the formation of a drop of solder at the bottom of the solder slot (Figure 1-22).

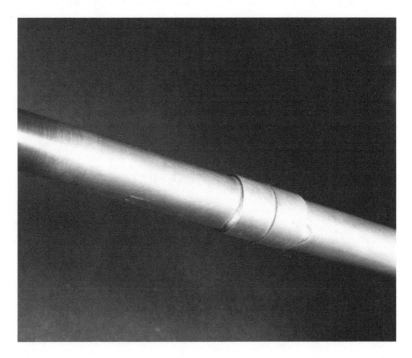

Figure 1-22. Photograph showing solder drop at bottom of solder slot in a horizontal coupling.

The appearance of a solder drop indicates the clearance volume was filled to overflowing. The addition of any more solder causes the drop to fall free.

When making a horizontal solder joint it is best to start to feed the solder into the upper half of the solder slot while observing the bottom half for the formation of an overflow drop.

When a joint is made with the solder slot facing downward a full solder ring may be recognized by the appearance of a solder fillet around the solder slot. The fillet is formed by the excess solder fed into the clearance volume after it is completely filled. This solder can no longer be held by capillary action and bulges out at the bottom where surface tension holds it against the force of gravity. The addition of any more solder results in the formation of a drop of solder at the lowest point in the fillet (Figure 1-23).

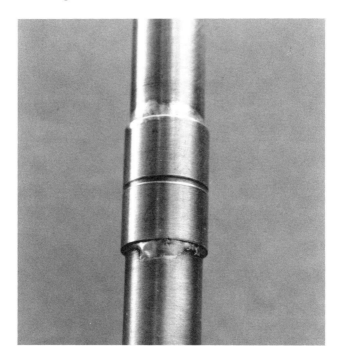

Figure 1-23. Photograph depicting a downward facing solder slot with solder fillet and solder drop formed after clearance volume is completely filled with solder.

The appearance of both a fillet and a drop is an indication that the clearance volume is full of solder which in turn indicates a good joint. Adding extra solder in this instance causes the fillet and drop to run off the connection.

When a joint is made with the solder slot facing upward no special indicators are usually present. In this case, a full solder ring is recognized by a solder band completely filling the solder slot and extending, in part, to the tube and rim of the solder fitting as illustrated in Figure 1-24.

Figure 1-24. A full solder ring in an upward facing solder slot.

Contrast the ring shown in Figure 1-24 with that in Figure 1-25. The latter shows an incomplete solder ring as evidenced by the clearance volume which is not completely filled with solder.

In soldering a joint that faces upward, it is best to feed solder in approximately a 120° arc on one side of the solder slot and to watch for the appearance of solder on the opposite side. As soon as solder wells up on the opposite side, filling the soldering slot, the job is complete.

Some service personnel prefer to add a solder fillet as the final step in soldering a joint. A fillet is unnecessary when the clearance volume is completely filled, as evidenced by a full solder ring. However, there is nothing wrong in adding a fillet—provided the heat is localized in the vicinity of the solder slot to produce the fillet without causing any solder to run out of the clearance volume.

Figure 1-25. Photograph showing a partially formed solder ring indicating that the clearance volume is not filled with solder.

Do not confuse the appearance of a barely visible, fine line solder ring with a full solder ring. When solder first begins to melt, it streaks out along a path of least wetting resistance determined by heat, clearance space and surface cleanliness. In many cases the initial path of least resistance occurs along the solder slot. If the solder pursues this path, a fine solder line appears before the solder is actually drawn into the clearance space. The solder is usually then drawn into the clearance space at a zone remote from the solder source. If additional solder is not fed into the solder slot, the joint may have a large unsoldered area as is shown in Figure 1-26.

The solder was fed into the solder slot at a point just above the unsoldered area in Figure 1-26. The solder streaked around the solder slot forming a fine line solder ring, entered the clearance volume on the side opposite of feed point (front side in the photograph), moved around the bottom of the joint and began to fill up the rear side when the solder ran out. The addition of supplemental solder would have enabled the joint to fill-up properly resulting in a full solder ring.

Figure 1-26. Photograph depicting the cup-end of a joint with a fine line solder ring and a large unsoldered area caused by insufficient solder.

Figure 1-27 is an enlarged view of the tube-end removed from the solder cup displayed in the previous illustration which shows the unsoldered area.

There are probably many soldered connections having flaws similar to those depicted in Figures 1-26 and -27 yet they still hold up well in actual field installations. Surprisingly, soldering is not a critical operation. A small amount of well-placed solder goes a long way in metal joining. With complete solder rings on top and bottom (and most of the clearance volume filled with solder) a flawed joint such as those shown in Figures 1-26 and -27 will withstand all but the toughest vibration and temperature conditions encountered in refrigeration applications.

Figure 1-27. The tube-end removed from the solder cup.

Trapped Flux Pockets

A problem related to solder streaking is the occasional formation of bare spots in the clearance space caused by trapped flux pockets. This phenomenon is most often encountered in soldering the larger sizes of soft copper tubing which have been slightly dented or bent out-of-round. The streaking solder surrounds and traps a pocket of flux which cannot be displaced by the solder. In some cases the flux thickens or becomes crystallized, sticks to the copper and leaves small bare spots which are void of solder (Figure 1-28). These flux pockets (bare spots) are usually too small to appreciably weaken the joint and can be eliminated by merely rotating or reciprocating the tube in the fitting while the solder is still molten.

Figure 1-28. Bare spots in soldered joint caused by trapped flux pockets.

Bare spots in a soldered joint can also result from poorly cleaned metal surfaces. A localized oily spot can defeat solder wetting action and leave a solder void in its place. A dented or out-of-round tube or fitting can also provide a localized clearance space, which is too great to promote effective capillary action, that results in a solder void.

Solder Run-off

Once a full solder ring has appeared, no more solder should be added since it will merely run out of the joint. Two situations are of particular interest:
- downward facing solder slots
- horizontal or upward facing solder slots

If the solder slot faces downward, excess solder fed into the slot simply drops out of the joint and runs over onto the copper below. Although no harm is done, this makes for an unsightly connection.

If the solder slot is horizontal or faces upward, excess solder may run into the tubing. Globs and slivers of solder travelling around inside the system can interfere with refrigerant flow control valves, plug strainers, and even work into and damage the compressor. A heavy glob of solder can also get trapped in a vertical run of tubing and generate rattling noises during compressor operation.

Although not much of a concern in refrigeration work, the problem of lead in water is receiving considerable attention. Some authorities believe the lead in 50/50 solder, commonly used in residential plumbing, works its way into the drinking water. The source of the lead is believed to be long slivers of solder which have been permitted to run into the tubing as a result of sloppy soldering procedures. This is just one more reason why it is important to practice proper soldering procedures and prevent solder from running into tubing.

The best way to prevent molten solder from running into the tubing is:
- learn how to recognize when a full solder ring has been achieved

- learn to estimate the amount of solder required to make a joint of a given diameter

For example, after soldering enough ½-inch OD tubing joints, you should develop an accurate notion as to approximately how much solder is required for each joint — as measured by the length of wire solder used. Any solder in excess of that amount will simply run out of the joint. Consequently, when soldering a horizontally or upward facing joint, carefully observe the amount of wire solder used. If more than the estimated amount is fed into the solder slot, it can only mean the excess has run into the tubing.

As a general guideline, when using standard .125-inch diameter wire solder, the length of solder used should approximately equal the diameter of the fitting. For example, in soldering a ¾-inch fitting, a piece of solder about ¾-inch long should be fed into the solder slot.

Overheating

Overheating is one of the principle causes of defective soldered joints. Soldering involves a number of chemical reactions which best take place within a narrow range of temperature and time. As previously stated, optimum soldering temperature is about 550°F , and the soldering operation should be completed as quickly as possible. There is no practical way to measure joint temperature while soldering, but if the solder flows into the soldering slot as soon as it reaches melting temperature, optimum temperature conditions have been attained.

If the joint is heated above optimum soldering temperature, a poor connection can result from three factors which act either separately or in combination:

1. Overheating interferes with proper fluxing action that takes place in a relatively narrow temperature range. When exceeding the upper limits of this temperature range, the flux loses its ability to clear away the oxide layer. As a result, the flux cannot prevent the oxide layer from reforming.

2. Overheating of the molten solder in the clearance space of the joint changes the chemical nature of the solder-copper bond. Examination of an overheated joint will reveal a mottled surface displaying a varied texture with a duller coloration than the original surface. New solder does not readily wet this mottled surface. In case of an overheated joint, it is best to remove the old solder (down to the base metal) and to start the procedure over again.

3. Overheating of the copper tube changes its metallurgical properties. Copper used in the manufacture of tubing must be oxygen-free. Even trace amounts of oxygen amounting to no more than 0.01 to 0.08% can weaken the copper when heated. Hydrogen, naturally occurring in the atmosphere, penetrates the copper at prolonged, elevated temperatures and then tends to combine with the oxygen to form water vapor. The water vapor which is formed becomes trapped in the copper—generating high pressures, and creates small cellular voids which greatly weaken the metal. This condition is known as hydrogen embrittlement.

Oxygen in the copper can also form traces of cuprous oxide which may collect in the grain boundaries further weakening the copper.

Copper producers incur the added expense of manufacturing oxygen-free copper which will not weakened due to chemical changes that may occur during soldering and brazing. Still, prolonged overheating can result in the copper losing its oxygen-free characteristics at the surface level. A copper surface damaged by hydrogen embrittlement and/or by impurity deposits in the grain boundaries will not be wet by solder as readily as an undamaged surface. Although these conditions do not occur frequently in soldering, they are noteworthy and should be briefly presented as possible explanations for those situations where difficulty is encountered in wetting an apparently clean copper surface.

SOLDERS

Solders used in the refrigeration trade fall into three categories:
- tin-lead
- tin-antimony
- tin-silver

The tin-lead solders are not recommended for use in refrigeration piping. Although successfully used for many years by the plumbing trade to join copper water pipe, they lack the creep resistance and strength characteristics required for the higher pressures, temperatures and vibrations encountered in refrigeration work.

The term "creep" is the slow displacement of metal under tensile stress. Different metals and soldering alloys have different resistance to creep. The creep of aluminum electrical wire under screw terminals was a cause of failure in the early use of aluminum wire.

Creep can permit the copper tube to be gradually pushed out of a soldered joint socket by pressure and vibration. The process can take a few days or many years, depending upon the temperature and pressure. A form of rapid creep can occasionally be witnessed during a copper water pipe freeze-up in the winter. As the freezing pressure builds up in the pipe, the copper tube can be pushed right out of the soldered joint exposing the soldered end.

The tin-lead family of solders is low in creep strength particularly at elevated temperatures that are typically found in the compressor discharge line. There is a general industry consensus that tin-lead solder should not be used in refrigeration piping. While it is true that thousands of tin-lead soldered joints have been in use for many years in refrigeration piping, most of these joints are located on lines not subjected to excessive vibrations and/or temperatures. Tin-lead solders should not be used in refrigeration piping applications since unpredictable changes may occur in the stress level of a joint and because it is not practical to use different solders on the same job. This convention should be adhered to at all times, and the use of tin-lead solder should primarily be confined to water and drain line joints.

Tin-antimony solder (95/5) has a high resistance to creep and withstands higher operating temperatures than tin-lead solders. Tin-antimony solder has been successfully used on copper refrigeration piping for many years. The only problem with 95/5 is that it is not recommended for use on brass valves and fittings. The antimony in the solder interacts with the zinc constituent of the brass to cause joint embrittlement.

Tin-silver solders are preferred for general refrigeration work. By alloying a small amount of silver with the tin, a solder with superior creep, vibration and temperature resistance is obtained. This solder is also compatible with brass valves and fittings in that no embrittlement results.

Tin-silver solder is not new. Prior to World War II its superior qualities were recognized, and this type of solder was referred to as "instrument solder." Due to the silver content, its relatively high cost precluded its use as a piping solder. However, experience gained during World War II with marine condensers and engine oil coolers proved the added cost of using tin-silver was justified in terms of reliability.

Tin-silver solder in wire form became available to the refrigeration trade shortly after the end of World War II. Consequently, with some 40 years of reliable field use behind it, its superiority as a solder for refrigeration work is well established.

Solder alloys other than lead, antimony and silver are currently available for use on copper and brass, but they have not come into general use in the refrigeration trade because of a reluctance on the part of service personnel to experiment with new materials when the "old reliables" have served so well.

Factors beyond the control of the refrigeration industry will likely force a change. The sudden concern voiced by some about the use of tin-lead solders for potable water is already causing changes in the domestic plumbing trade, and new alloys are coming into use.

In June of 1986, President Reagan signed amendments to the Safe Drinking Water Act. These amendments require the use of "lead-free" pipe, solder and flux in the installation or repair of any public water system. Under the provisions of these amendments, solders and flux are considered lead-free when they contain not more than 0.2% lead. Pipes and pipe fittings are considered lead-free when they contain not more than 8.0% lead. These requirements went into effect immediately upon the President's endorsement. However, the law gives state governments until June of 1988 to implement and enforce these new limitations. A number of states have already banned all use of lead materials in drinking water systems.

Plumbing contractors engaged in air conditioning work might find it convenient and economical to use the same solders in both trades. At the same token, economic factors are also encouraging the reduction or elimination of silver bearing alloys. It is likely that solders formulated without lead and of reduced silver content will fast gain acceptance due to the recent amendments. For example, Engelhard markets SILVABRITE 100 which is an alloy of tin, copper and silver. The Harris company markets STAY-SAFE 50 which is an alloy of tin, silver, zinc and antimony. Other manufacturers will no doubt also follow suit and introduce their own brands of lead-free solders to do their share in complying with the requirements set down by the Safe Drinking Water Act.

Special Solder Properties

In addition to the strength and creep resistant properties described previously, solders exhibit various properties that necessarily result in different handling characteristics. Although not much of a concern in soldering copper tubing, these handling characteristics play an important role in selecting solders for electrical, sheet metal and automatic machine soldering.

The reasons solder differences play a minor role in copper tube soldering are:

- The trade has narrowed down the selection to 4 or 5 solders — each having similar handling characteristics.

- By using a hand held torch, a degree of heat control is provided that can compensate for any of the solder differences that may exist.

Still, for those with a broader interest in the subject, some supplemental information on solders follows.

No pure metal possesses all the properties necessary to make a good solder. Pure lead melts at 621°F which is too high for most soldering. Pure tin melts at a desirable 449°F but lacks the required strength. Through trial and error ancient metal workers discovered that the best solders resulted from mixing two or more metals to form an alloy. An alloy can be described as a metallurgical mixture of two or more metals that in combination have physical and chemical properties different from either of the constituent metals.

When tin is mixed with lead, for example, the melting temperature of the lead drops and the resulting alloy takes on its own properties—primarily dependent upon the relative proportions of tin and lead. A picture of what occurs as an alloy melts and solidifies can be seen on what the metallurgist calls a *constitutional diagram*. Refer to Figure 1-29.

The vertical lines on the chart represent the percentages of tin (Sn) in the alloy. The extreme right hand vertical line represents 100% tin; while the extreme left hand line represents 100% lead (Pb). The horizontal lines represent temperature as indicated in degrees Fahrenheit on the sides of the chart.

The chart is divided into three distinct zones, each representing the three different states of the alloy. The first zone, indicated by the 45 degree hatching, represents the solid state. The second zone indicated by the dashed lines represents the liquid state. The third zone indicated by the circles and dash lines represents the pasty state where liquid and solid particles are mixed together.

The three zones on the chart are separated by the liquidus and solidus lines. The line A-E-B separating the liquid and pasty zones is known as the liquidus line. All points on the chart above the liquidus line indicate a completely molten state.

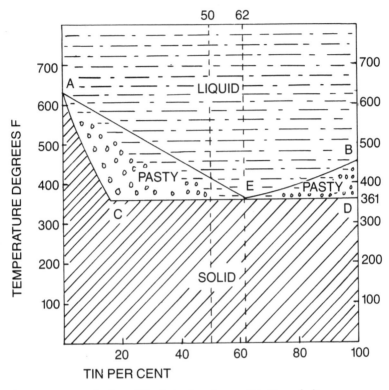

Figure 1-29. A simplified tin-lead constitutional diagram.

The line A-C-E-D, separating the solid and pasty zones, is known as the solidus line. All points on the chart below the solidus line indicate a completely solid state. The area between the liquidus and solidus lines defines points on the chart where molten liquid and solid particles coexist in various proportions.

The explanation of the constitutional diagram begins with point E. Note that this point is the only point on the diagram common to the solidus and liquidus lines. Moving along the dashed vertical line through point E, the diagram reads 62% tin. The alloy at point E is made up of 62% tin and 38% lead.

Moving laterally across the diagram, the alloy at point E is seen to melt at 361°F. This is also the lowest melting point of any mixture of lead and tin and is known as the "eutectic solder." Another feature of the eutectic solder is that it is the only alloy composition which transforms directly from the liquid state without passing through the pasty state.

The common 50/50 solder is diagramed next. Entering the base line at the 50% tin point and moving vertically along the dotted line, the solidus line is intersected at the 361°F point and the liquidus line at the 421°F point. This means that as a 50/50 solder is heated it starts to melt at 361°F and is not completely molten until it reaches a temperature of 421°F. In the process it passes through a pasty range of $(421° - 361°)$ $= 60°$ where the solid and liquid solder coexist before completely liquefying. The pasty range can be compared to the slush formed in a brine before it freezes solid.

The pasty range adds the feature of moldability not found in the eutectic solder. In certain applications, the ability to reposition or build up a joint with solder before it sets is a desirable feature. Examining the diagram shows that as the tin content is reduced to the 20% level the pasty range increases while the liquidus temperature increases. On the other hand, if minimal soldering temperature is a primary concern, a eutectic solder would be best suited for the application.

The constitutional diagram yields much useful information pertaining to solder melting temperatures and physical states. It does not, however, give any information pertaining to solder strength, wettability or flow characteristics that are necessary in matching solder to a particular application. In the final analysis, this determination has to be made through trial and error. However this task need not fall to service personnel. The craft of copper tube soldering has fortunately become so well advanced over the years that little or no such experimentation is actually required for selecting the right solder for a given job. Rather, the detailed research and development manufacturers have devoted to producing soldering alloys makes intelligible selection of the proper solder a relatively straightforward task.

SEPARATING SOLDERED JOINTS

Fitting and Tube Seizing

Surprisingly, considerable difficulty may be encountered in taking apart a soldered connection, particularly with tube sizes greater than ⅝-inch OD. The difficulty exists due to the creation of tin-copper compounds, most commonly Cu_6Sn_5, at the interface between the copper and solder. These compounds are hard and brittle, and have surfaces that do not readily slide over one another (Figure 1-30).

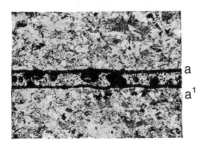

Figure 1-30. Magnified section through a soldered joint (a and a^1 show tin-copper compound layers sandwiching a layer of solder).

As long as the solder is molten it continues to react with the copper to produce thicker layers of tin-copper compounds. This explains why it is difficult to separate a soldered joint that has been overheated or kept in the molten state for excessively long periods of time. Due to that fact, it is a valid conclusion that speed is just as essential in separating a soldered joint as it is in making the joint. The longer the solder remains in the molten state, the more copper it combines with and the harder it is to separate.

Solders rich in tin, such as those used in refrigeration work, are more prone to generate tin-copper compounds than solders rich in lead. Obviously that is why joints soldered with 50/50 tin-lead solder are generally easier to separate than joints soldered with tin-silver or tin-antimony solders. Conversely, joints which are easiest to unsolder are weaker in creep strength. This is one of the reasons why tin-lead solders are not recommended for refrigeration work.

Additional factors which may complicate the separation of a joint should be considered.

1. **Tube size.** The larger the tube size, the greater the force necessary to separate the joint. It takes more force to take apart a 1⅜-inch coupling than a ⅝-inch coupling. A 1 ⅜-inch coupling has a much larger surface area and proportionally more force is required to shear the solder bond across the tin-copper interface layers.

2. **Joint clearance.** The tighter the clearance space between the tube and fitting, the greater the force necessary to take the joint apart. By increasing the clearance space, the tin-copper interface layers are separated more readily and the joint can be pulled apart by shearing the sandwiched solder layer (see Figure 1-30). This fact can be readily verified by attempting the field unsoldering of a swaged or a tube-within-a-tube joint. The added clearance in these joints always makes them easier to separate than a standard sweat fitting with tightly controlled clearance space.

Since the ability to take a joint apart is of relatively minor importance in comparison to joint strength, joint clearance should not be permitted to exceed .006-inch. Allowing more clearance merely to enable easy unsoldering should not become common practice.

3. **Out-of-roundness.** A soldered connection can be bent out-of-round during installation or when servicing. For obvious reasons, such a joint will present mechanical obstruction problems as well as unsoldering problems during disassembly.

Breaking the Connection

The following is one procedure recommended for separating soldered joints. As a preliminary step, make sure all system internal pressure has been relieved and that the line to be unsoldered is open to the atmosphere. Note: <u>molten solder blown out of a joint by internal pressure can injure personnel</u>.

First, spread some paste flux around the joint. When melted, the flux acts as a lubricant, particularly if the joint shows some indication of seizing while being disassembled.

As soon as the solder melts, strike the joint a sharp separating blow with a mallet or a piece of wood. If the joint being separated is an elbow, strike the elbow bend. If it is a straight coupling, an elbow or bend of some type can usually be found in the vicinity of the joint being taken apart to receive the separating blow (Figure 1-31).

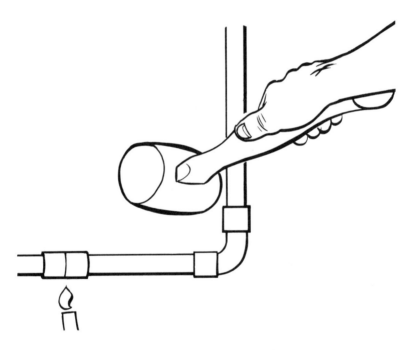

Figure 1-31. Separating a soldered coupling.

If there is no elbow or bend in the vicinity, the joint can be separated by pulling on a section of tubing remote from the heated fitting.

Removing a Fitting

A fitting can be driven off a tube-end with a piece of hard wood that has a squared-off end. After the solder becomes molten, the fitting should be struck by sliding the piece of wood along the tubing so it collides with the lip of the fitting end (Figure 1-32).

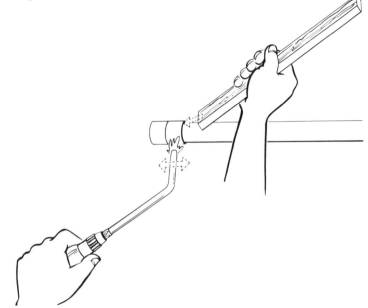

Figure 1-32. Drive a fitting off the tube-end using with a piece of wood.

In more difficult situations where a fitting appears to be seizing up, grasping the fitting with pump pliers and rocking it back and forth should help to loosen the fitting for removal.

When using this procedure, do not grip the fitting too tightly because it can be bent out-of-round. This fitting removal procedure is related to that used when removing an automobile brake drum which is being held-fast by the brake shoes. In some situations a slight rotary rocking motion applied in addition to the lateral rocking motion described can be advantageous in removal of the fitting.

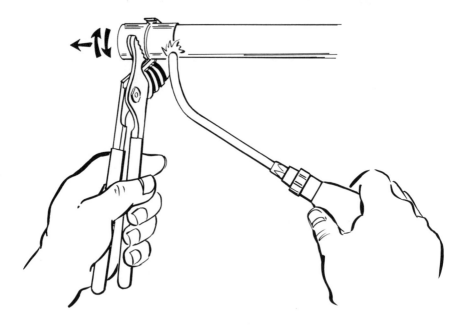

Figure 1-33. Removing a fitting by "walking" it off the tube-end.

Heat Control

In any attempt to separate soldered joints, it is important to control the heat application. Raising the temperature beyond what is necessary to keep the solder in a molten condition is not recommended. By exceeding the melting temperature the joint will only become more difficult to take apart. In fact, overheating can cause a soldered joint to permanently seize and make it impossible to separate. The only recourse in removing a permanently seized fitting is to cut it off.

EQUIPMENT AND TECHNIQUES

Fuel Cases

Acetylene

Acetylene is the oldest and still the most common gaseous fuel used in the refrigeration trade. Acetylene is a compound of carbon and hydrogen (C_2H_2) and is lighter than air. But as an unstable compound, it cannot be pressurized above 15 psi in a gaseous state without exploding. In order for acetylene to be stored in a cylinder, it is dissolved in acetone which in turn is held in a porous material, calcium silicate, that completely fills the cylinder.

Acetylene cylinders must be handled with care. They cannot be dropped, dented or subjected to any rough handling. As a safety precaution acetylene cylinders must always be maintained in an upright position. If laid on the side, when the cylinder valve is leaking or is not positively closed, the acetone can leak out leaving behind a dangerous build-up of acetylene pressure.

Although acetylene is a colorless gas, it has a distinctive garlic-like odor. Never operate an acetylene torch in the presence of an acetylene gas odor. Check for leaks at all connections with a soap solution or refrigerant leak detection solution.

The most popular cylinder size in the refrigeration trade is the B tank followed by the much smaller MC tank. The B tank is about 23-inches high with a 6-inch diameter. It holds about 40-cubic feet of acetylene. The MC tank is 14-inches high with a 4-inch diameter and a capacity of 10-cubic feet of acetylene (Figure 1-34).

The B tank coupled to a regulator and torch handle of the type shown in Figure 1-35 is the standard heat source for the refrigeration trade. A wide variety of tips are available that can interchangeably be screwed into the handle. With a torch temperature of 4,000°F and tip sizes designed to suit just about any application, this air-acetylene outfit can handle the majority of the soldering and brazing needs encountered in refrigeration work.

MC   **B**

Figure 1-34. MC and B acetylene tanks (Courtesy, L-TEC Welding & Cutting Systems).

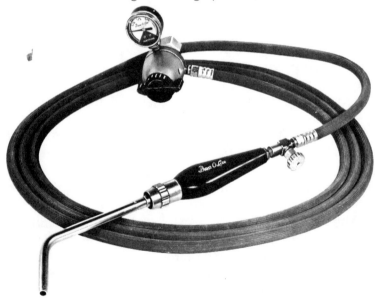

Figure 1-35. Regulator, hose, handle and tip for B size tanks (Courtesy, L-TEC Welding & Cutting Systems).

Special tips are also available which swirl the mixed air and acetylene at a very high velocity through a propeller-like set of vanes at the rear of the flame tube. This swirling action produces a higher energy and relatively short flame that concentrates heat output on the work precisely where it is required.

A refrigerant leak detector, as shown in Figure 1-36, is also available that can be attached to the handle instead of a tip. Because of the intense, sharply focused flame, acetylene heated leak detectors are the most sensitive and easiest to use of all leak detectors which employ an open flame.

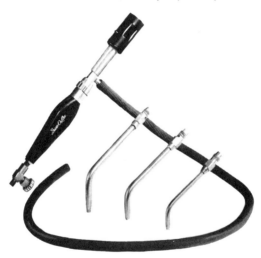

Figure 1-36. Leak detector used with air-acetylene outfit (Courtesy, L-TEC Welding & Cutting Systems).

Oxy-Acetylene

When combined with oxygen, acetylene burns at a temperature near 6,000°F enhancing its usefulness for cutting and welding steel.

The most popular oxy-acetylene outfit utilized in the refrigeration trade is a portable rig that combines an MC acetylene tank with an R oxygen cylinder (Figure 1-37). This combination is light enough to be carried up a ladder onto a

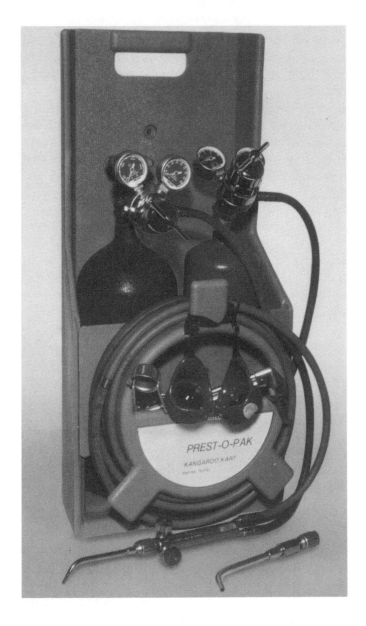

Figure 1-37. Hand carried oxy-acetylene unit (Courtesy, L-TEC Welding & Cutting Systems).

roof when necessary. An optional adapter is available that converts the handle for use with air-acetylene to help conserve oxygen when higher temperatures are not required. Although air-acetylene is adequate for most tube brazing, oxy-acetylene proves superior in many instances. For example, in brazing the suction and discharge lines to the stub tubes on a compressor shell, the ability of the oxy-acetylene torch to sharply focus a high temperature flame at the precise location of the connection greatly reduces the possibility of weakening the connection of the stub tube to the compressor shell.

The ability to braze or weld a broken fan bracket or a compressor mounting bolt greatly enhances the value of this unit making it almost indispensable in the practice of refrigeration and air conditioning service and installation.

Larger oxy-acetylene outfits using acetylene and oxygen tanks of larger capacities are also available on wheeled carts for shop use. Some of these units are equipped for both oxy-acetylene and air-acetylene service. An extra outlet on the acetylene regulator, shown in Figure 1-38, is used to feed a separate hose with an air-acetylene torch.

The chief advantage in using acetylene is that it is the most versatile of all the fuel gases. It burns at the highest temperature making it the only fuel gas which can be used for welding, cutting, brazing as well as for soldering. Another important advantage for acetylene is that it has a high burning velocity. This means that combustion takes place at the tip end in a tightly confined heating zone rather than in a long, wide, dancing flame.

The ability of a torch to develop a concentrated heat zone at the tip end is referred to as "flame focusing." Since a concentrated flame is best for the soldering and brazing of copper tubing, the sharply focused flame of acetylene lends it an advantage in this regard.

The disadvantage in using acetylene is that it does not come in disposable containers. Instead, acetylene requires a heavier tank which must be refilled.

Figure 1-38. Oxy-acetylene and air-acetylene unit mounted on hand truck (Courtesy, Uniweld Products, Inc.).

Propane

Propane is a gas used widely as fuel for both home and industry. Recovered during the process of distilling crude oil, propane is sold in a liquefied form in a wide variety of containers. The common term for propane is LPG (liquefied petroleum gas) or LP-gas.

In air, propane burns at a temperature of approximately 3,300°F. It is heavier than air and, as with all fuel gases, should only be used in adequately ventilated areas.

The advantages in using propane are its low cost and ready availability. The common 20-lb propane tank has long been popular in the plumbing trade and is gaining popularity in the refrigeration trade as well. Used in both trades, this same tank is also used with back yard grills and on recreation vehicles. The convenience of getting the common propane tank refilled also seems to be helping make the use of propane more wide spread.

When equipped with a regulator, a length of hose, a handle and selection of tips (as shown in Figure 1-39) most soldering and brazing operations called for can be accomplished with this basic set-up. In addition, the tank is

Figure 1-39. A 20-lb propane tank equipped with a soldering and brazing outfit.

useful for such large heating jobs as thawing out pipes and raising the temperature of cold coils to assist in evacuating a system.

The chief disadvantage in using propane is that it burns at the lowest temperature of all the fuel gases. This can create some problems, particularly with the tougher brazing jobs which require higher temperatures.

To increase the effectiveness of propane, swirling-mixing tips should be used instead of standard tips. These tips have internal baffling that creates a swirling-mixing of the gas and air and ultimately generates a hotter, more concentrated flame. Swirling-mixing tips can be identified by the enlarged, stainless steel mixing chamber on the tube-end as displayed in Figure 1-40.

Swirling-mixing tips allow very little flame adjustment. In fact, each tip must be operated at a rated minimum pressure. Flame adjustment is made by selecting the tip size rated for the job. Manufacturers typically supply up to seven different swirling-mixing tips to cover the flame range required for most applications.

Figure 1-40. Propane burner tip with swirling combustion chamber.

Another tool available which can increase the effectiveness of propane is the amplifier shield. This is a curved piece of stainless steel that clips onto the torch tip to redirect and reflect heat back at the work area. The shield also protects neighboring components from heat damage (Figure 1-41).

By using swirling-mixing tips and amplifier shields, the use of propane can be effectively extended into the brazing range.

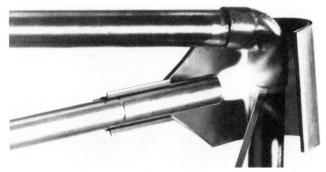

Figure 1-41. Stainless steel amplifier shield (Courtesy, Turbotorch).

While most manufacturers of soldering and brazing equipment offer swirling-mixing tips and heat shields, these components are not interchangeable between the various brand names on the market. The shields must be matched to the tips, and the tips must be matched to the specific handle produced by a given manufacturer.

Another critical point to remember is that propane and acetylene equipment are not interchangeable. Equipment designed for acetylene must only be used with acetylene; and equipment designed for use with propane must only use propane or other approved propane type gas.

The most common propane tank is the 14.1-oz size commonly sold in hardware stores. Soldering tips that screw directly onto this compact tank providing a hand held tank torch (Figure 1-42) are readily available. Refillable propane tanks smaller than 20-lb capacity are also available. A refillable tank holding 5-lb of propane is shown in Figure 1-43.

Swirling-mixing tips as well as leak detector attachments are also available for hand held torches. One problem with the older torches like these was flame drown-out. Whenever the tank was upended to access a difficult soldering position, the flame was extinguished by liquid propane that seeped out. This problem has been solved in the newer torches by means of liquid fuel checks that have been built into the tips. These compact tank torches can now be operated in just about any position without causing drown-out.

Another improvement is the swivel tip. This allows the

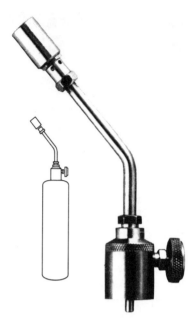

Figure 1-42. Common hand held propane tank torch (Courtesy, Goss, Inc.).

Figure 1-43. A 5-lb refillable propane tank.

tank to be held upright while the tip is allowed to swivel to any position.

For service personnel, who operate out of the trunk of an automobile and specialize in service, the small 14.1-oz hand held propane tank torch is probably adequate. And it may suffice for their occasional soldering, light brazing and leak detection applications. However, in order to meet the more demanding requirements of commercial refrigeration and air conditioning service/installation, the propane torch must be backed up by a good oxy-propane outfit. Undoubtedly occasions will arise requiring the higher temperatures and better focused flame of the oxy-propane torch.

Propane can be mixed with oxygen to burn at a temperature of about 4,580°F. The most popular oxy-propane outfit in the trade combines the 14.1-oz propane tank with an R oxygen cylinder in a carrying stand (Figure 1-44).

The increased temperatures obtainable with the oxy-propane torch allows brazing, some cutting but no steel welding capabilities.

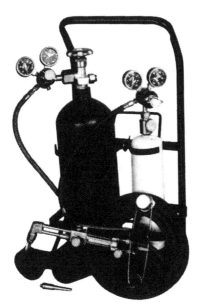

Figure 1-44. Portable oxy-propane outfit (Courtesy, Goss, Inc.).

Enriched LP Gas

The third, and most recent addition of fuel gas to the trades can be described as enriched LP-gas. This type of fuel has long been available in bulk form for industrial use. Some fuels are enriched natural gas or propane. Others are propylene or blends of various hydrocarbons. They all have the advantage of burning at a higher temperature with a more sharply focused flame than that of propane, although less than that of acetylene. These fuels can also be liquefied and stored in containers similar to propane tanks.

It is likely that enriched LP-gas will continue to grow in popularity as more suppliers market their gas and equipment.

MAPP Gas

MAPP gas was one of the earliest blended LP-gases to be used in the trades. It is a hydrocarbon blend that burns at a higher temperature than propane but has related handling characteristics otherwise. (MAPP is a registered trademark of the Air Reduction Company, Inc.)

In normal atmospheric conditions, MAPP gas burns at a temperature of approximately 3,400°F compared to about 3,300°F for propane and 4,000°F for acetylene. With oxygen, MAPP burns at a temperature of 5,300°F compared to 4,580°F

Figure 1-45. Hand held MAPP tank torch with swivel tip (Courtesy, Turbotorch).

for propane and 6,000°F for acetylene. The increased temperature and more sharply focused flame gives MAPP a decided advantage over propane. The disadvantages, it is more expensive and is not as readily available as propane.

MAPP gas is sold in 16-oz disposable containers which can be fitted with a regulator and swivel tip so it can be used as a hand held tank torch (Figure 1-45).

The 16-oz container can also be fitted with a regulator, length of hose, handle, and a number of tips to provide a general soldering and brazing outfit (Figure 1-46).

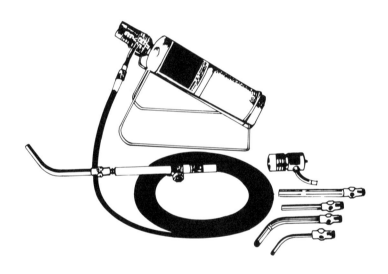

Figure 1-46. MAPP soldering and brazing outfit.

MAPP gas is also available in larger tanks. A 5-lb MAPP tank with regulator and hose attached is shown in Figure 1-47.

MAPP gas can be mixed with oxygen to burn at a temperature of approximately 5,300°F. As with propane, the most popular Oxy-MAPP outfit in the trade combines the 16-oz MAPP tank with an R oxygen cylinder rigged in a carrying stand (Figure 1-48).

Figure 1-47. A 5-lb MAPP tank with attached soldering and brazing outfit.

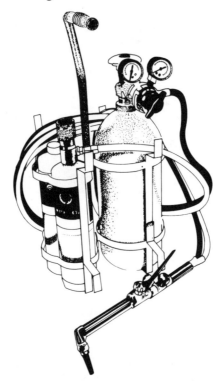

Figure 1-48. Portable Oxy-MAPP outfit.

FG-2 Gas

Another enriched LP-gas used in the trades is FG-2 marketed by Prest-O-Lite. FG-2 has flame properties similar to MAPP but is reported to work more effectively in cold weather applications. At a temperature of 32°F , FG-2 generates a tank pressure of 87 psig as compared to 68 psig for MAPP gas and 63 psig for propane.

FG-2 is supplied in a 15.2-oz disposable tank designated DF. A regulator and tip are available and can be screwed directly onto the DF tank to provide a hand-held tank torch (Figure 1-49). This torch can be used for soldering as well as brazing and can be operated in any position. An added advantage it that the regulator and tip also fit disposable MAPP and propane tanks.

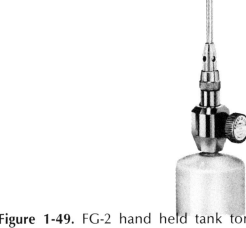

Figure 1-49. FG-2 hand held tank torch (Courtesy, L-TEC Welding & Cutting Systems).

For remote operation, a regulator, 12-feet of hose, and a torch handle are available which can be coupled to the disposable tank (Figure 1-50). This assembly is most convenient in making joints in cramped quarters where adequate room is not available to manipulate the hand-held tank torch.

Figure 1-50. Disposable DF tank with attached soldering and brazing outfit .

Figure 1-51. A refillable FB tank with regulator, hose and handle (Courtesy, L-TEC Welding and Cutting Systems).

Should a larger gas supply be required, the refillable FB tank is available (Figure 1-51). This tank will hold 7 ½-lbs of FG-2 gas, and its squat, stable shape is patterned after the popular 20-lb refillable propane tank. The tank has its own hand operated shut-off valve which receives an LPG regulator. When connected to a length of hose and a torch handle, this tank has the capacity to feed the largest tips normally used in the trades. The extra capacity of this tank also makes it useful in thawing frozen pipes or in raising the temperature of an evaporator to assist in vacuum dehydration.

Manufacturers of soldering and brazing equipment are marketing a line of tips which are common for propane, MAPP and FG-2 use as well as most of the other enriched LP-gases. The tip orifices are designed to provide satisfactory operation over a range of slightly different gas characteristics. However, some of the older tips and even some of the newer tips are still not interchangeable. These tips are specifically designed for propane or MAPP and are not recommended for use on any other gas. Until the situation regarding propane and enriched LP-gases is standardized, it is the responsibility of the operator to make sure the tips employed are suitably matched to the gas being used.

Finally, the choice of which fuel gas or torch set-up to use depends entirely upon individual circumstances and the personal preferences of the operator. For example, the contractor engaged in refrigeration and air conditioning installation might find the versatility of oxy-acetylene preferable. On the other end of the scale, the appliance dealer who primarily services domestic refrigerators and freezers might find the MAPP or FG-2 tank torch adequate. Somewhere in the middle ground between the two might be the plumbing, heating and air conditioning contractor who may find the 20-lb propane tank with assorted tips most satisfactory for his work. The advantages and disadvantages of the major fuel gases have been enumerated, yet because at present no concrete standards exist, it is difficult to make specific recommendations concerning which gas or set-up is best suited for any given circumstance.

Wrought Copper Fittings

Both ACR soft copper tubing supplied in rolls and hard-drawn copper tubing that comes in straight lengths are joined by wrought copper sweat fittings. The same fittings are used for soldering and brazing. In addition to providing the necessary clearance for capillary action, the fittings must not be porous under the action of the high pressure refrigerant. For this reason wrought copper sweat fittings are preferred over cast fittings for refrigeration applications. The drawing and stamping process involved in manufacturing wrought fittings provides a dense metal that is impervious to gas leakage.

Wrought copper fittings are available in a wide variety of styles and shapes to fit all sizes of ACR tubing. The principle shapes are illustrated in Figure 1-52.

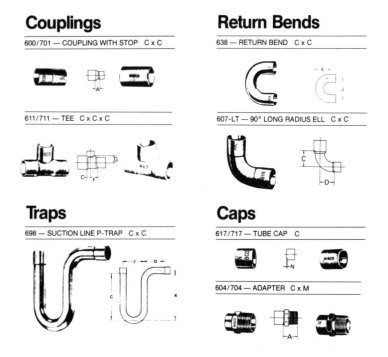

Couplings

600/701 — COUPLING WITH STOP   C x C

611/711 — TEE   C x C x C

Traps

698 — SUCTION LINE P-TRAP   C x C

Return Bends

638 — RETURN BEND   C x C

607-LT — 90° LONG RADIUS ELL   C x C

Caps

617/717 — TUBE CAP   C

604/704 — ADAPTER   C x M

Figure 1-52. Common wrought copper sweat fittings (Courtesy, NIBCO, Inc.).

Swaged Connections

When joining tubes of the same diameter, a swaged connection can be used instead of a straight sweat coupling. In a swaged connection a tube-end is expanded by a specially designed swaging tool to receive the end of the other tube with a clearance gap of about .005-inch. Swaged connections may be soldered or brazed. The swaging operation is performed using swaging tools illustrated in Figure 1-53.

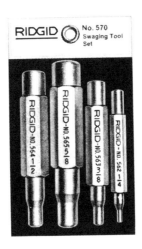

Figure 1-53. Swaging tool set (Courtesy, Rigid Tool Company).

ACR soft and hard-drawn copper tubing in the popular ¼-, ⅜-, ½- and ⅝-inch sizes are commonly swaged using a matching swaging tool. A section of the tubing is clamped in a flaring block with enough tubing extending from the top to receive the guide and expanding portions of the swaging tool. While holding the flaring block in one hand, the swaging tool is hammered down into the top end of the tubing until it has been driven up to the collar or hilt of the swaging tool as demonstrated in Figure 1-54.

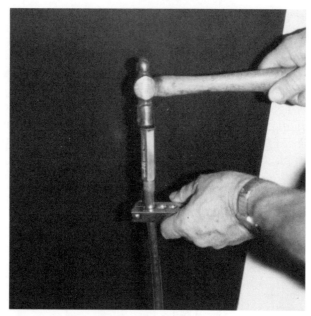

Figure 1-54. Making a swaged connection with a swaging

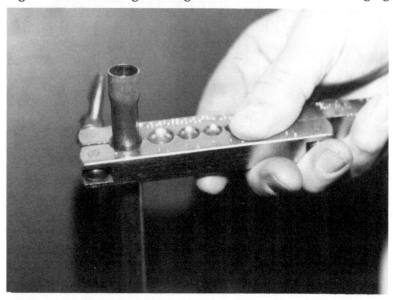

Figure 1-55. Swaged tube end shown clamped in flaring block.

Figure 1-55 shows the swaged tube-end after the swaging tool has been removed.

A single tool (Figure 1-56) with all the expanding sections machined on one surface may be advantageous to include in a tool kit rather than having to carry an entire set of swaging tools for various tubing sizes.

Figure 1-56. One piece swaging tool for common ACR tubing up to ⅝-inch (Courtesy, Ritchie Eng. Company).

The punch-type swaging operation shown in Figure 1-57 has been mechanized in a combined flaring and swaging tool. With this tool, quick change swaging adapters can be installed under the hand driven screw to swage the tube-ends. Swaging adapters are available to swage the common tubing sizes from 3/16- to ⅝-inch.

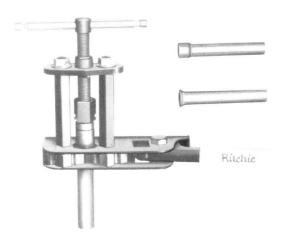

Figure 1-57. Combined flaring and swaging tool (Courtesy, Ritchie Eng. Company).

Tube Expander

Another way of joining tubes of the same diameter without sweat couplings is with the tube expander.

In a tube expander, a set of expandable jaws is inserted within the tube-end and then cammed outward to enlarge the tube diameter by the action of a lever. The expansion is made large enough to receive an unexpanded tube-end while still providing the proper clearance for soldering and brazing.

Each tube size requires a separate die head which is installed on the tool. The tool shown in Figure 1-58 can handle tube sizes from 5/16-inch up to 1 5/8-inch when supplied with a complete set of heads. This sort of expanding tool can only be used on annealed tubing. When working with hard-drawn copper tubing, the tube-end which is to be expanded must first be annealed. This can be done simply by heating the tube-end to a dull red and then cooling it with a damp rag.

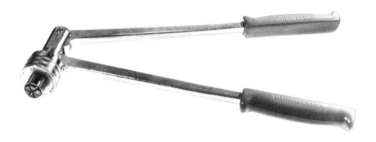

Figure 1-58. Single action tube expander (Courtesy, Ritchie Eng. Company).

There is an advantage in using the tube expander over manual swaging. The score lines which are usually left on the tubing below the swage by the clamping or flaring block can set up zones where fatigue cracking may occur if the tubing is subjected to extremely severe vibration. This problem can be avoided by swaging with an expander tool which requires no clamping or hammering and, as a result, leaves no score lines.

The T-DRILL system (illustrated in Figure 1-59), which is related to swaging, allows branching into a main line without the need for a tee-fitting. It also serves to make distribution manifolds on the job. In the T-DRILL system a uniquely designed drill is centered on the main line at the coupling point. When energized, the drill advances and forms a pilot hole in the line. Once inside the line, formation cams are released which are drawn upward to form a brazing collar which receives the branch tube.

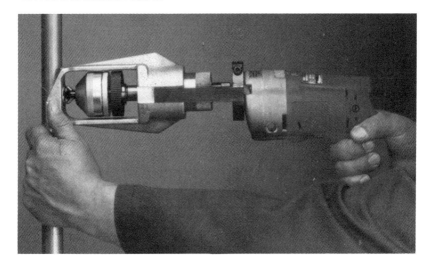

Figure 1-59. T-DRILL for branching into a main line (Courtesy, T-DRILL).

GENERAL SOLDERING TIPS

- **Use only ACR tube for refrigeration work.** Basically, copper tube is available to the mechanical trades in two principal classes:

    Plumbing tube — types K, L, M and DWV
    ACR tube — air conditioning and refrigeration work
    All types of copper tubing are available in hard-drawn straight lengths. Types K, L and ACR are also available in soft-drawn coils. The plumbing types are designated in nominal sizes which are always ⅛-inch diameter less than the actual size. For example, a copper tube in the plumbing trade which is designated ½-inch will actually be ⅜-inch diameter. ACR tube is always designated by its actual diameter. For example, a ½-inch ACR tube is actually ½-inch OD.

    All tube used in the refrigeration trade must be of the ACR type because it is specially manufactured to eliminate moisture and contaminants. It also has a wall thickness designed to handle the pressures encountered in refrigeration work. Conversely, plumbing types M and DWV have a lighter wall thickness. ACR tube is also designated *dehydrated* or "manufactured to meet requirements ASTM B 280." It is always sealed or capped. ACR tube should be purchased from a reputable refrigeration wholesaler thereby ensuring the buyer that it is actually ACR tube and not tube designated for plumbing applications.

    The dimensions of ACR tube are given in Table 1 of the Appendix.

- **Do not solder a tube-within-a-tube connection without a reducing coupling.** Copper tubes of different size should never be joined by inserting one within the other and bridging the gap with molten solder. In the field, this type of connection can occasionally be observed for increasing the size of a liquid or suction line. Taking the case of a ¼-inch tube inserted within a ⅜-inch tube, the clearance gap would be about .035-inch. This is far in excess of the .002-

to .006-inch clearance recommended for proper strength. It also exceeds the .01-inch clearance limit where capillary action is effective in drawing in the solder. The strength of this type of connection depends on the solder used, which is the weakest link. Vibration and stress can eventually cause solder to crack and leak. The cost of reducing couplings is so nominal that they should always be used in soldering copper tubes of different diameters.

- **Never solder a system under pressure.** Even at relatively low system pressures of only 1 or 2 psi, it is not possible to make an effective soldered joint. The force of capillary action which draws the solder into the clearance space is so slight that it cannot resist any internal pressure originating from the system. If the internal pressure is great enough, it will simply blow the solder out of the clearance space. And if the internal pressure is much weaker in nature, it can still form fine *blow-holes* through the molten solder. Recall the hazards involved in attempting to solder a joint under system pressure.

When repairing a leak in an operating refrigeration system it is necessary to discharge the pressure and leave the system open to the atmosphere until the soldered connection has been completed. The point where the system is exposed to atmosphere should be located as close to the soldered connection as possible in order to eliminate any back pressure build-up.

When repairing a leaky soldered joint, some service personnel simply discharge the refrigerant into the atmosphere and then reclose the service valve believing the pressure has been completely relieved. But in fact, pressure immediately starts to build-up again as the remaining gas begins to warm and as the refrigerant oil releases its dissolved gas. It is imperative that this "reserve" pressure be vented by cracking the service valve and keeping it open until the solder repair is completed.

- **Never solder a line containing refrigeration oil**. Refrigeration oil circulates with the gas and can collect at low spots and bends in the piping system. When repairing a leak in an existing system, be aware of the presence of oil—particularly in low spots. Oil can interfere with the soldering operation by preventing solder wetting action. The best way to remove the oil from the line is to vent the gas then cut the line near the leak so gravitational forces can drain the oil. The draining operation can be facilitated by bending the severed lines to the lowest possible position. Some tapping and shaking of the lines can also speed up the draining process. As soon as the oil is removed, clean and resolder the joint before any new oil can accumulate.

- **Preliminary tinning strengthens joints**. Tinning is the process of separately coating the joint surfaces with a wetting layer of solder. By tinning the surfaces before making the joint, the quality of the fluxing and wetting action can easily be observed. In the preliminary tinning of a soldered tubular joint the end of the tubing is coated with solder and the excess is wiped away using a clean, dry rag of natural fiber. The inside of the fitting is coated in a similar manner except that the excess is merely shaken off the inner surface. The joint is then lightly fluxed, assembled and heated using just a *touch* of extra solder—if necessary.

    In normal soldering, a joint does not have to be pre-tinned. This practice requires a lot of extra time to perform and, consequently, is not recommended on a regular basis. Basically it is a useful practice for those hard-to-reach situations where proper visual inspection is often limited or may be hampered altogether.

- **Do not mix solders**. Different solders should not be mixed when soldering the same joint. For example, if a 50/50 solder joint were to be resoldered with tin-silver solder, the lead in the solder used originally can alloy with the tin-silver reducing its creep and vibration resistance. In addition, the "hybrid solder" will take on new, unfamiliar properties that are different from either of the original solders. And the resulting handling and flow characteristics the hybrid demonstrates may not be what the operator expected from the tin-silver solder. A similar result occurs when lead contaminates tin-antimony solder (95/5).

When remaking a previously soldered connection, and the solder type is unknown, it is best to start all over again. Begin by melting the old solder out of the fitting and off the end of the tube. The fitting can best be cleared by holding it in a pliers and heating it to the point where the solder becomes molten. Then with a swift, jerking downward motion "throw" the molten solder out of the fitting and onto the floor. The excess solder on the tube-end can be removed by heating the solder to its molten state and wiping it off with a clean, dry rag as illustrated in Figure 1-60. A natural fiber wiping rag such as cotton is best suited for this purpose.

After removing the bulk of the solder, the wetting film in the fitting and on the tube-end should be cleaned down to the base metal by means of wire brush and abrasive cloth. The joint is then ready for resoldering.

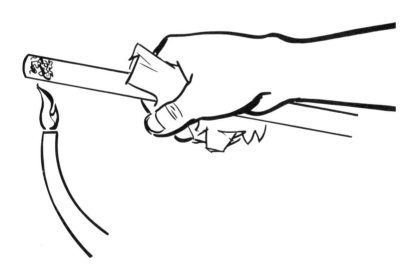

Figure 1-60. Molten solder should be wiped off tube-end.

- **Solder layering**. A technique occasionally employed in the plumbing trade to solder copper tube in sizes greater than 2-inches involves layering two different solders that have two different melting temperatures. The technique is used when difficulty is experienced in preventing solder from running out of a large sweat joint. The joint clearance may be too large resulting in the hydrostatic pressure of the liquid solder exceeding the force of capillary action. And because of the difficulty in uniformly heating a large joint, some local overheating may occur promoting solder run-out. To overcome these problems, plumbers apply a higher melting solder (such as 95/5 tin-silver or 95/5 tin-antimony) to form a partial dam at the bottom of the joint. The remaining clearance space is then filled with solder such as 50/50 tin-lead which has a lower melting point. Careful control of the heat is necessary to draw the 50/50 solder into the joint without melting the higher melting point solder.

- **Pinhole leaks in brazed joints can be repaired by soldering**. Although it is preferred practice to repair a brazed joint by rebrazing, special situations may arise making soldering a more feasible alternative. Suppose a pinhole leak develops in a brazed joint after a number of years of service. Assume the joint is located in a congested area where the higher temperature required for brazing could damage some adjacent plastic pipe or electric cable.

  A simplified repair can be effected by first locating the pinhole leak with a bubble solution. Thoroughly clean the area surrounding the leak using abrasive cloth or wire brushes to bring the metal to a high shine. After venting any internal line pressure, flux and solder the cleaned area. Withdraw the torch at short intervals, and rub the end of the wire solder on the metal in the area of the pinhole to make sure the surface is completely tinned. As the final step, build up the area above the pinhole leak with an extra layer of solder for extra strength. Tin-silver solders are best for this type of repair.

SOLDER or BRAZE?

Equipment and compressor manufacturers generally recommend brazing over soldering for general metal joining when used in conjunction with their products. Some manufacturers even go so far as to specifically exclude soldering as a viable practice, but this recommendation may be extreme. A definite and important role for soldering exists in the refrigeration trade, and consequently, its role is increasing as more service personnel discover the advantages of soldering over brazing.

On the assembly line, in a manufacturing plant, brazed joints are definitely preferred over soldered joints. The availability and convenience of oxy-acetylene torches along with a nitrogen sweep to eliminate copper oxide formation makes for clean, tight joints at the compressor, the condenser and all interconnecting piping. But contrast this assembly line situation with a field service call where a serviceman might be required to climb a ladder at night to repair a leak in the liquid line of a rooftop unit. According to recommended procedures, the serviceman would have to carry an oxy-acetylene rig and a tank of high pressure nitrogen gas to the roof, connect a nitrogen sweep and then rebraze the joint. The practicality of this sort of situation is such that it is rarely done. The simplest solution — repair the leak with a sweat fitting using tin-silver solder and a hand-held tank torch.

Refrigeration service is generally done under emergency situations, or at best, under extreme pressure. The task of the serviceman attempting to handle a substantial number of calls during a long, hot summer can be grueling. The serviceman needs all the help he can get by way of alternate approaches to any and all service procedures — including soldering and brazing. Consider some specific examples.

Suppose a serviceman is replacing a 3-ton hermetic compressor in a residential air conditioning system. While preparing to braze the suction and discharge lines into the stub tubes of the compressor, the serviceman discovers that his oxygen bottle is empty. Even though a back-up hand held tank torch might be in the truck, the smaller and cooler flame of this torch does not adequately lend itself to rapid brazing. The brazing can only be accomplished at the compressor stub tubes using a maximum flame over a long period until the joint is heated up to the required temperature. However, this extended interval of heating may very well overheat the compressor shell and increase the chances of additional heat damage. Under the circumstances, rather than risk damaging the unit, the best solution is to solder the joints using a tin-silver solder and a moderate LP-gas flame.

A related example is a leak that has occurred in close proximity to a condenser or an evaporator. Suppose the serviceman only has an air-acetylene outfit with a small tip. The heat from the torch would be drawn away too rapidly by the finned surfaces to bring the joint up to brazing temperature. Here again, a sweat connection using tin-silver solder is the best answer.

On larger commercial refrigeration and air conditioning systems, with suction lines greater than a 1 ⅝-inch OD, it is much easier to install and remove suction line filter driers by soldering than by brazing.

Soldering can also be advantageous when installing all types of brass valves into a system. Many of these valves contain soft seats, packings and power elements which can be damaged by heat. Even though they may be fitted with extended stub tubes and the precaution of wrapping them in wet rags has been observed, these valves are still often damaged during brazing. The much lower temperatures incurred in soldering offers a distinct advantage in this situation.

Under carefully controlled conditions, soldering may be employed to attach components to a refrigeration line under pressure. On many of the older domestic refrigerators the capillary tube is soldered in a heat exchange relationship with the suction line. Due to long term effects of vibration and chemical attack, in many cases a soldered bond has the tendency to separate. The repair procedure called for under these circumstances consists of resoldering the capillary tube to the suction line while the system is under pressure. This procedure has been practiced in domestic refrigeration service for many years apparently without any harmful effects on the internal chemistry of the system.

The preceding are just a few of the many service situations where soldering is the solution to difficult field repairs.

The primary type of metal joining a serviceman uses also tends to determine what kind of equipment must be carried in a service vehicle. The small independent contractor still plays a major role in the refrigeration trade. To keep operating expenses down he typically operates out of a small truck, a mini-van, a station wagon or even out of the trunk of an ordinary automobile.

At the very least the serviceman must carry three tanks of refrigerant (R12, R22, and R502) for basic servicing. If commercial installation work is involved, he must also carry an oxy-acetylene rig, an air-acetylene torch as a back-up, and a nitrogen tank. This all adds up to a considerable load of high pressure gas containers. Transporting this equipment not only constitutes a personal hazard, but it takes up a good share of cargo room. If the work load is basically service, with an occasional residential air conditioning unit installation, the serviceman may only need to carry an LP torch for soldering or, instead, an air-acetylene outfit for soldering and light brazing. Carrying only what may be required can eliminate the need to unnecessarily transport high pressure oxygen and nitrogen tanks. On the other hand, if the serviceman intends to do commercial refrigeration and air conditioning installations, there are no short cuts. A full complement of brazing and soldering equipment may prove necessary.

In conclusion, the importance of soldering in the air conditioning and refrigeration trade cannot be over-emphasized. The advantages soldering offers has proven time and again to be invaluable to the serviceman who knows and practices correct soldering techniques and can recognize when they are properly applied.

PART II _____

BRAZING

INTRODUCTION

The refrigeration/air conditioning industry is the largest worldwide consumer of silver containing brazing filler metal. Millions of pounds are supplied annually for the manufacture of heat transfer coils and for the field installation of refrigeration and air conditioning equipment.

The principle use of silver brazing filler metals is in the joining of copper tube connections on return bends in evaporators and condensers and in the fabrication of copper interconnecting piping. Occasionally some brass and steel piping connections are required, particularly at compressor service valves and flow control devices.

Brazing filler metals used in refrigeration applications must possess excellent joint strength, flexibility and corrosion resistance to withstand the temperature, pressure, vibration and corrosive atmospheres encountered. In addition, the need to maintain internal system cleanliness and the need to avoid damaging costly system components imposes special constraints on refrigeration brazing which are typically not called for in non-refrigeration applications.

Repair statistics have shown that the principal cause of refrigeration system failure is loss of refrigerant due to piping leaks. Many of these failures could have been prevented with improved dissemination of the latest information on preferred soldering and brazing procedures. The payback in savings of refrigerant and perishable product load is so great that all indications point to a need for increased emphasis on proper soldering and brazing instruction in refrigeration training programs.

Most books on brazing do not recognize the special problems associated with the refrigeration trade and treat the subject in an incidental manner along with general tube brazing. Manufacturers of brazing filler metals do, however, publish information directed primarily to the refrigeration trade. Designed primarily as vehicles to promote a manufacturer's line of filler metals, these pubications, while informative to a certain degree, are usually abbreviated pamphlet formats that are not very comprehensive.

Because of the dominant role played by the refrigeration industry both in the amount of brazing filler metal consumed annually and in the shear number of brazing operators it employs, refrigeration brazing deserves independent recognition in brazing literature. Consequently, the primary objective of this portion of the text is to assemble and present a comprehensive yet practical guide—lending both theory and actual technique, to brazing as it relates to the special rigors of the refrigeration trade. And because refrigeration brazing requires special considerations well above and beyond that necessary in general brazing applications, the instruction and techniques included in this text should serve to foster better overall metal joining techniques of a more generalized nature.

TERMINOLOGY

Brazing and soldering, though related by many common features are, nevertheless, significantly different metal joining processes. The higher temperatures required by brazing changes the chemistry of the operation as compared to soldering. In brazing, copper is raised to a temperature above 1,000°F where annealing and chemical changes occur. Filler metals containing phosphorus require no flux when used on copper. Soldering, on the other hand, is accomplished at a temperature well below that generating any significant physical or chemical changes, and flux is always required in open-air soldering.

An oxy-acetylene or other oxy-fuel torch is frequently used in brazing. The oxygen and fuel ratio in these torches must be adjusted to produce a reducing flame. Soldering is performed with air-acetylene or other air-fuel gas torches which have fixed air to fuel ratios.

Brazing produces tougher joints than soldering. Brazed joints consequently withstand higher temperatures and are more creep and vibration resistant. Brazing also allows for a much looser fit in connections which yields the additional advantage of facilitating emergency repairs in broken piping and in improvising non-standard connections when the need arises.

Because brazing is greater in scope than soldering, the brazing process requires more working experience and a greater knowledge of filler metals, torches and heating techniques. These are just a few of the differences between brazing and soldering which will be addressed in greater detail. Refrigeration service personnel are often called upon to make and repair soldered and brazed connections and must subsequently acquire a proficiency in both metal joining techniques.

The American Welding Society defines brazing as: "a group of welding processes which produces coalescence of materials by heating them to a suitable temperature and by using a filler metal having a liquidus above 840°F and below the solidus of the base materials. The filler metal is distributed between the closely fitted surfaces of the joint by capillary attraction."

Consequently, it is safe to conclude that capillary attraction (action) is a critical factor in the formal definition of brazing. Figure 2-1 shows a typical brazed joint. A copper reducing fitting is shown with the tube-end (spigot) inserted within the socket or cup of the fitting. The fitting is then heated to brazing temperature. Assisted by capillary action, brazing filler metal is drawn into the clearance space between the socket and spigot to complete the assembly.

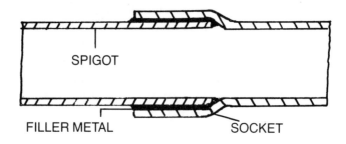

SPIGOT

FILLER METAL   SOCKET

Figure 2-1. Cut-away illustration of a typical capillary brazed joint.

There are a relatively small but important number of brazed connections employed in the trades which do not depend upon capillary action. These connections are generally used in system modifications or to effect emergency repairs. A few examples are given below.

Figure 2-2 shows a ¼-inch OD tube brazed into a ¾-inch OD suction line. To effect the connection, a ¼-inch hole is drilled in the wall of the suction line and the ¼-inch tube inserted within the hole to extend slightly below the wall. A fillet is brazed around the base of the ¼-inch tube to seal it to the suction line. Such connections are commonly found in refrigeration systems to install a Schrader valve. Another use is to connect an equalizer line from the evaporator outlet to the expansion valve.

Figure 2-3 illustrates a field assembled distribution manifold for an evaporator or a condenser. A main line is pierced with a series of spaced holes to receive the branch lines. The assembly is completed by brazing fillets around the branch lines.

Figure 2-4 represents a 1 ⅛-inch OD suction line which developed a pinhole leak by rubbing against a neighboring electrical conduit. A simple repair may be effected by filling the depression with a glob of brazing filler metal making sure the glob wets the copper base metal.

Brazed connections of the type not utilizing capillary action are termed "braze welding" by The American Welding Society to distinguish them from brazed connections requiring capillary action.

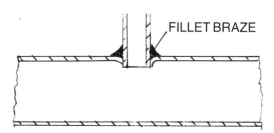

Figure 2-2. Cut-away illustrating the positon of a fillet (relative to the brazing mouth) which lends support to the small vertically positioned tube.

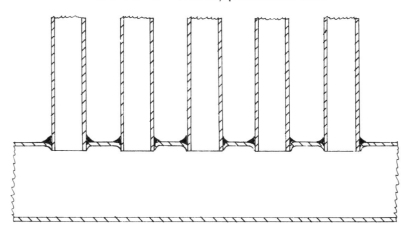

Figure 2-3. Illustration of a typical refrigeration manifold construction employing multiple brazed connections which demonstrates the use of fillet brazing.

Figure 2-4. Illustration showing use of a brazing glob to seal a pinhole leak.

Since a trend has developed in the refrigeration trade to draw a sharp distinction between welding and brazing, the use of the term "braze welding" is a source of confusion. The term brazing is used in the refrigeration trade when referring to all brazed connections—whether capillary action is involved or not. This discourse adheres to that convention. However, there is a distinction between the two classes of brazed joints. For the purpose of discussion in this text, the term "fillet brazing" is used to describe a brazed joint requiring an external fillet as illustrated in Figures 2-2, -3 and -4.

The term "capillary brazing" is used in reference to the brazed connection that requires capillary action to draw the filler metal into the clearance space.

BRAZING FILLER METALS

Working Properties and Composition

Brazing filler metal (or brazing alloy as it is commonly referred to in the trades) is a bonding alloy which melts above 840°F but below the melting temperature of the base metals which are to be joined. The alloy is formulated to wet the base metal and penetrate the surface to form a continuous metallurgical bond. Wetting of the base metal by the molten filler metal is the most essential requirement for effective brazing. Without wetting, the brazed joint may temporarily hold due to a superficial mechanical bond, but is sure to fail eventually.

Wetting can be explained by first examining the two extremes in making bonded tubing connections. Figure 2-5 shows a typical socket and spigot connection sealed with a thermosetting resin material such as epoxy. The uncured plastic is easily applied at room temperature and allowed to harden. The temperature of the tubing is not raised to

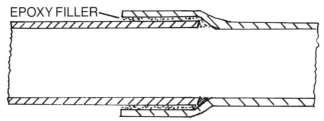

Figure 2-5. Socket and spigot connection sealed with epoxy filler.

cause any annealing or chemical changes. The procedure is simple and has been occasionally employed to repair tubing connections for aluminum condensers and aluminum interconnecting tubing on automobile air conditioning systems. Unfortunately, the plastic to metal bond is only mechanical in nature and depends upon the plastic locking onto the surface roughness in the metal. A combination of vibration, elevated temperature and chemical deterioration with time, may cause the joint to fail. Because of its high failure rate, this type of connection has not found favor.

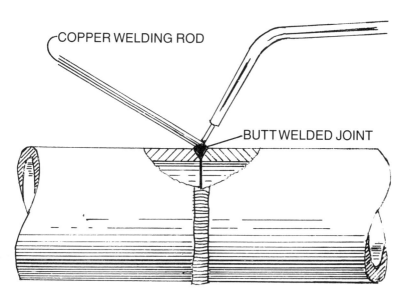

Figure 2-6. Butt welded copper pipe connection.

At the other extreme, Figure 2-6 shows a butt-welded copper pipe connection. A copper welding rod deposits molten copper onto the joint which mixes with some molten copper from adjoining pipes to form a solid copper connection. Welding produces the strongest joint and is the preferred joining method where conditions permit its use. Obviously, thin walled tubing cannot be conveniently welded in the field.

What is required for thin walled tubing is a joining process that falls between the plastic bond of Figure 2-5 and the molten metal bond of Figure 2-6. Brazing, using filler metals which wet the base metal, fulfills that need.

Base Metal Wetting

Wetting of the base metal by molten filler metal is explained with reference to Figure 2-7 which shows a test setup employed by the metallurgist to conduct wetting tests. A 1 ½-inch square metal plate about ¹⁄₁₆-inch thick and formed of the base metal to be tested is supported above a source of heat. The top surface of the plate is carefully cleaned and fluxed. The plate is heated until the flux melts at which time a carefully weighed sample of filler metal is dropped on the center of the plate. As the filler metal melts, it begins to spread over a portion of the plate. An infrared sensor is employed to measure the temperature at which the filler metal begins to melt (solidus) and the temperature when it becomes completely molten (liquidus). The time taken for each event is also recorded. After the filler metal is completely molten, the plate is allowed to cool and the area of the spread is measured. The spread area and the spreading time are measures of the wettability of each filler metal sample in relation to the particular base metal.

Assume the plate in Figure 2-7 is copper and the filler metal is an alloy containing 15% silver, 80% copper and 5% phosphorus, designated as BCuP-5 by the American Welding Society. When the plate reaches a temperature of 1,190°F the alloy begins to melt (solidus) and will be completely molten

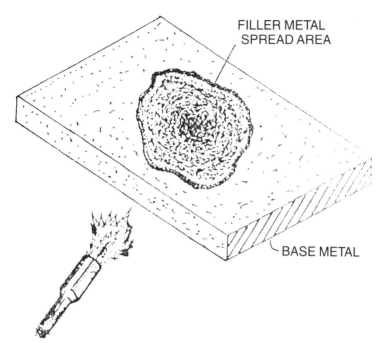

FILLER METAL
SPREAD AREA

BASE METAL

Figure 2-7. Test for determining filler metal effectiveness.

at a temperature of 1,475°F (liquidus). This particular alloy spreads rapidly and covers a relatively wide area indicating its effectiveness on copper.

The plate is then sectioned through the spread area and examined under a high power microscope. Photomicrographs of the area are taken which reveal the changes in crystalline structure at the interface between the alloy and base metal. A schematic representation of such a photomicrograph is shown in Figure 2-8. The molten filler metal penetrates the copper a

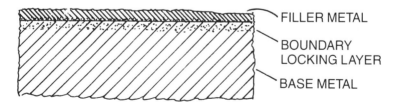

FILLER METAL

BOUNDARY
LOCKING LAYER

BASE METAL

Figure 2-8. Enlarged view illustrating how boundary locking layer serves as transition between filler metal and base metal.

few thousandths of an inch in depth creating a boundary layer which interlocks metallurgically with the copper. The system now has three distinct layers:

- the copper base metal
- the BCuP-5 filler metal layer
- the boundary locking layer

The boundary locking layer, which interfaces the copper base metal and the filler metal layer, consists of a new alloy formed mostly by phosphorus penetration of the copper crystals. The three layer system forms a stack of continuous interlocking crystals.

When the filler metal is employed to bond two overlapping plates, the system can be represented as in Figure 2-9—which is essentially Figure 2-8 in a sandwiched relationship. Figure 2-10 is a cross sectional view illustrating wetting action in a typical brazed copper coupling connection. The molten filler metal is drawn into the clearance space by capillary action where it attacks and penetrates the overlapped copper surfaces forming two opposed boundary locking layers.

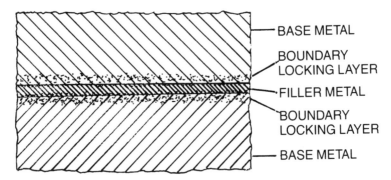

BASE METAL

BOUNDARY LOCKING LAYER

FILLER METAL

BOUNDARY LOCKING LAYER

BASE METAL

Figure 2-9. View similar to Figure 2-8 illustrating a lap joint.

Figure 2-11 is a greatly enlarged view of the boxed portion of Figure 2-10 schematically illustrating the intergranular penetration of the base metal by the filler metal. Actual photomicrographs are different, but Figure 2-11 illustrates how two overlapping copper sections are joined by a process which may be termed "alloy welding." Instead of melting copper sections together as in welding, a similar result is

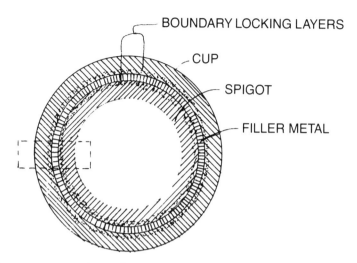

Figure 2-10. Enlarged cross sectional view of brazed coupling illustrating action of boundary locking layers.

achieved by alloy penetration of the copper. A single continuous piece of metal is achieved consisting of interlocking crystals of copper (moving from left to right in Figure 2-11), boundary layer alloy, brazing alloy, boundary layer alloy and copper. Strength tests show that the brazed joint is as strong as the copper itself when the proper brazing alloy is used. The selection of brazing alloys for best results is covered in the ensuing section on brazing alloy applications.

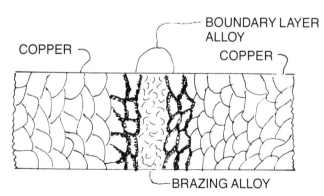

Figure 2-11. Schematic reproduction illustrating continuous crystalline structure across brazing bond.

As indicated, wetting is the most important requirement for good brazing. Although the principles of wetting are also broadly applicable to soldering, wetting as it relates specifically to brazing involves many more complicating factors. These factors will enter into a subsequent discussion on capillary action, heating and brazing alloy compositions.

CAPILLARY ACTION, CLEARANCE SPACE

The role of capillary action and clearance space as explained in connection with the soldering process also applies to brazing. The same fittings and swaging tools which provide a .002- to .006-inch clearance space for soldering are also used for brazing. When the clearance space is in this specified range, maximum penetration of the joint by the brazing alloy takes place to produce the strongest connection.

The emphasis on a tightly controlled clearance space in refrigeration tubing brazing is a carryover from general brazing where clearance space may be critical in providing a strong bond. For example, Figure 2-12 shows a shaft formed by two, 2-inch diameter steel rods brazed together by a butt joint. To obtain a strong bond, the opposing surfaces must be spaced with a capillary opening between them. Otherwise the brazing alloy will not be completely drawn into the clearance space. In those areas where the capillary space is exceeded a void is left behind which weakens the joint. The precise capillary spacing can be as low as .002-inch, depending upon the material being brazed and the brazing alloy used. In large ferrous and nonferrous structures the capillary spacing is critical in order for the brazing alloy to be draw into the far reaches of the opposed surfaces.

While the same guidelines for clearance space set out in general brazing work should be followed in refrigeration tube brazing, the practical realities of field service and instal-

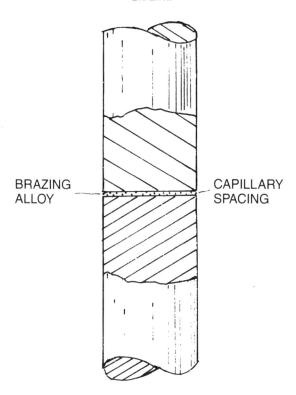

BRAZING
ALLOY

CAPILLARY
SPACING

Figure 2-12. Steel rods joined by a butt joint requiring critical capillary spacing.

lation often require a departure from suggested procedures. To begin, copper tubing and fittings can easily be damaged in normal handling. Annealed copper tubing in sizes greater than ½-inch is subject to being bent out-of-round in the normal process of uncoiling and stringing into place. And a pronounced distortion occurs at sharp tubing bends, even when made with a bending spring or tube bender. At joint locations where the tubing or fitting is distorted, the clearance space—in part, may greatly exceed the specified maximum of .006-inch.

There is also the matter of the necessarily large inventory of sweat fittings a service vehicle may be required to carry. Because of the wide range of tubing sizes and the large complement of fittings which go with each size, it is difficult to maintain a complete inventory at all times. Field situations continuously arise when the desired fitting is not available. In this instance, improvising a substitute connection on the job may be necessary.

Finally, there is the relatively small but important number of brazed connections (such as shown in Figures 2-2, -3 and -4) made by fillet brazing which do not depend upon capillary action.

The ability of brazing to form dependable tube connections where the capillary space is exceeded or does not exist at all, is an important advantage of brazing over soldering. Loosely fitted joints which would fail when soldered can be made permanent by brazing.

This brazing ability is derived from three principle properties:

- **braze spreading**—the ability of brazing alloy to spread over the base metal
- **braze shaping**—the ability of brazing alloy to be shaped into position
- **braze strengthening**—the ability of brazing alloy to add strength to the connection because of its relatively high tensile strength.

Braze Spreading

Braze spreading will be explained in connection with the test set-up shown in Figure 2-13. A 6-inch piece of ½-inch OD copper tubing is thoroughly cleaned to remove surface oxides. An indentation is placed at the center of the tube with a dull punch as a reference point. The tube is then fluxed and placed vertically in a vise. The indentation is then heated by the flame of an air-acetylene torch. As soon as the copper heats up to the proper soldering temperature, 50/50 tin-lead wire solder is fed into the indentation and the movement of the solder observed. Note that the solder runs straight down under the influence of gravity. Without the benefit of capillary action the solder will not move upward or sideways—only

96

Figure 2-13. Test setup showing no spreading in vertical soldering.

straight down. This illustrates that soldering relies on capillary action to move against gravity in order to fill the clearance space.

The experiment depicted in Figure 2-13 was repeated using a brazing flux and brazing alloy. The results are shown in Figure 2-14. Note that the brazing alloy spread all around the indentation, and was drawn up, down and sideways by the spreading action. This indicates that spreading operates independently of capillary action to move brazing alloy into brazing position.

The spreading experiment of Figures 2-13 and 2-14 has been conducted for nearly all the soldering and brazing filler metals used in the trades. In each case, the results were substantially the same. The solders were drawn straight down by gravity with no tendency to spread upward while all the brazing alloys demonstrated both a distinct upward and sideways spreading movement.

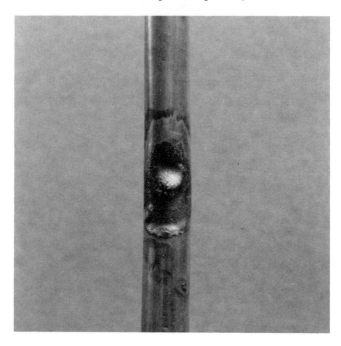

Figure 2-14. Test setup showing braze spreading.

While recognized in the brazing literature as a distinct phenomenon, the subject of spreading still often remains misunderstood for the most part. The extent of spreading varies with the base metal, and this aspect is particularly pronounced in copper. Spreading is also influenced by the action of flux which reduces the surface tension in a manner that promotes spreading. Braze spreading is less of a force in drawing brazing alloy than capillary action, but its importance in brazing is that it can supplement the weak force of capillary action in a loosely fitted joint.

Braze Shaping

The term braze shaping is employed here to describe a tacky range of operation exhibited by certain brazing alloys which enables them to be shaped into a desired form. Most brazing alloys used on refrigeration piping exhibit a tacky range of operation. Some alloys are designed not to have a tacky range so as to be very fluid in order to penetrate tight clearance spaces.

The popular brazing alloy designated BCuP-5 by the American Welding Society is an alloy composed of 15% silver, 80% copper and 5% phosphorus. It is sold by J.W.Harris Co. as STAY-SILV 15, by Engelhard as SILVALOY 15 and by Handy & Harmon as SIL-FOS. The alloy is generically referred to in the trade as "15% silver." The latter designation will be employed in this text. 15% silver has good braze shaping properties and can be used to braze loosely fitted joints which would otherwise fail if merely soldered.

Figure 2-15 shows a typical situation where a tube-end is inserted in an off-center fitting. Note the large gap on one side of the clearance space. Difficult piping situations of this type are encountered when remodeling or when moving equipment to a new location. An obstruction may deflect the tube just enough to prevent a straight junction.

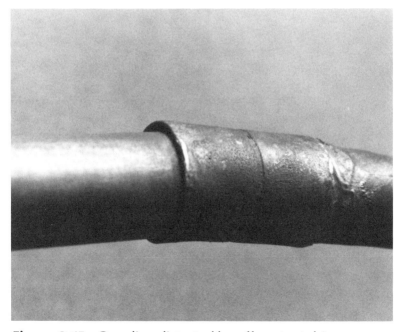

Figure 2-15. Coupling distorted by off center tubing.

Figure 2-16. Brazing alloy being pushed and troweled into large clearance gap.

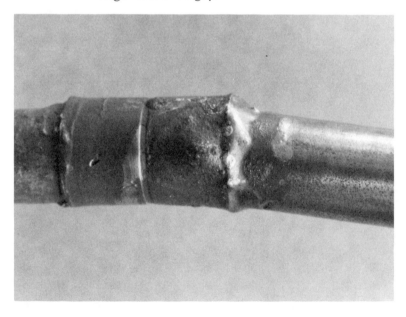

Figure 2-17. Brazed connection of Figure 2-16 complete with fillet.

The connection in Figure 2-15 may be effectively brazed using a brazing alloy such as 15% silver. After the joint is brought up to brazing temperature, the end of the brazing rod is inserted into the clearance gap. As the end becomes molten, the rod is used as a tool to push and trowel the brazing alloy into the gap until it is completely filled as is shown in Figure 2-16. A heavy fillet is then built up around the opening for added protection. The completed joint is displayed in Figure 2-17.

Braze Strengthening

The term braze strengthening is used here to convey the idea that the brazing alloys used on copper tubing are stronger than the tubing itself. Hard drawn copper tubing begins to anneal at about 700°F and is fully annealed at about 1,200°F. In the process the tensile strength drops from about 55,000 psi to about 32,000 psi. By way of comparison the tensile strength of 15% silver brazing alloy is about 85,000 psi. The high tensile strength of this brazing alloy can be used to compensate for the lack of any overlapped capillary brazed sections of base metal which provide the strength in a standard brazed tubing joint.

A braze strengthening application is shown in Figure 2-18 where a ¼-inch OD copper tube is fillet brazed into a ¾-inch OD suction line for connecting a Schrader valve. In the absence of any overlapped base metal, the brazing alloy itself is relied upon to strengthen the relatively weak copper. The large brazing alloy fillet effectively supports and bonds the ¼-inch line into place.

In pulling the above connection apart for testing, note how the ¼-inch copper line separates while the brazing alloy remains undisturbed as is shown in Figure 2-19. When attempting to saw through a fillet brazed joint with a hacksaw, note just how much harder it is to cut through the brazing alloy than the copper.

A connection such as shown in Figure 2-18 should never

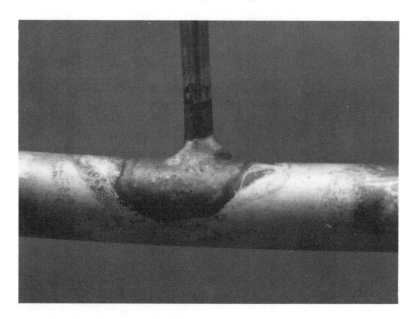

Figure 2-18. Brazing alloy fillet bonds tubing together.

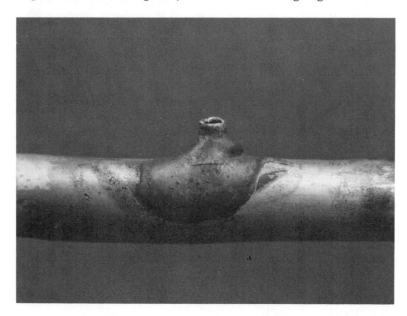

Figure 2-19. Destruct test showing failure in copper base metal.

be made with solder. The extremely low tensile strength of solder will not lend adequate support of the tubes if subjected to any rough handling.

Tube-Within-A-Tube Connection

Braze spreading, braze shaping and braze strengthening can act separately or together in a particular brazing application. All three act together in brazing the familiar tube-within-a-tube connection. Although not "approved" by standard brazing manuals because of excessive clearance, field experience indicates that this improvised connection holds up well in service. The discussion that follows only refers to the brazing process. As previously mentioned in the soldering discourse, this type of connection should never be made with solder.

The use of tube-within-a-tube connections originated due to the convenient fact that ACR tubing is supplied in common sizes ranging from ¼-inch OD to ⅞-inch OD, graduated in ⅛-inch increments. Figure 2-20, drawn from data in Table 1 of the Appendix, lists the relevant information.

Size	OD	Wall	ID	Clearance Space
¼	.250	.030	.190	¼ in ⅜ (.030)
⅜	.375	.032	.311	⅜ in ½ (.030)
½	.500	.032	.436	½ in ⅝ (.028)
⅝	.625	.032	.555	⅝ in ¾ (.020)
¾	.750	.042	.666	¾ in ⅞ (.018)
⅞	.875	.045	.785	

Figure 2-20. Optimum clearance space when one ACR tube is inserted in the next larger size.

Columns 1 and 2 list the tubing in nominal size and in decimal form, respectively. Column 3 lists the wall thickness. Column 4 lists the inside diameter. Column 5 lists the clearance between two adjacent tubes when the smaller is inserted into the next larger size. The clearance space is arrived at by subtracting the outside diameter of the smaller tube from the inside diameter of the larger tube and dividing by two. The clearance space so calculated serves as a rough guideline. In reality, the tubes are never concentrically arranged. When one tube is inserted within the other, the clearance space may vary along the insertion zone from zero to twice the listed clearance. This variation is illustrated in Figure 2-21 which shows a ⅜-inch OD tube inserted within a ½-inch OD tube.

Figure 2-21. Clearance space variation across insertion zone.

It is obvious that a tube-within-a-tube connection depends on factors which go beyond capillary action for its strength because of the wide clearance. An important factor is the additional insertion length. The end of the smaller tube is usually inserted within the larger one at least 1-inch, which exceeds the normal insertion length of a fitting socket. The strength of this type of connection is derived not only from the additional insertion length of the small tube but also from the extra brazing alloy required to fill the clearance space created. In most cases, 5- to 7-inches of .050 × ⅛-inch brazing rod is required. The preferred brazing alloy is a 15% silver or its equivalent.

In assembly, after cleaning and lightly fluxing the tube-ends, insert the smaller tube within the larger to a depth of about an inch. Mark off a 7-inch length of .050 × ⅛-inch 15% silver rod or equivalent with a punch to serve as an indicator of the amount of brazing alloy consumed. Next, heat the joint until it displays a dull, cherry-red coloration. Once heated sufficiently, feed the rod into the brazing mouth. Play the flame back-and-forth across the entire insertion zone, occasionally pausing at the far end in order to maintain the temperature higher there than at the mouth end. This effectively draws the brazing alloy into the deepest reaches of the joint.

Feed the first 3- or 4-inches of the brazing rod into the brazing mouth rather rapidly in an attempt to completely fill the clearance volume. When sufficiently filled, a fillet begins to form around the brazing mouth. As a check, allow the brazing mouth to cool and solidify. If a sinkhole develops in the mouth as shown in Figure 2-22, it is an indication that additional brazing alloy is required. Reheat and continue feeding brazing alloy into the mouth at a slower rate while carefully observing any tendency of the brazing alloy to build up around the brazing mouth. Because of the tacky nature of 15% silver or equivalent, it is held in the clearance volume by surface tension and does not run out of the joint until the clearance volume is completely filled. To complete the job, ring the mouth with a large fillet which serves as a backup for any internal weak spots that may exist within the joint. Note, if more than 6- or 7-inches of the rod is consumed by the brazing mouth, the procedure has not been accomplished properly.

Figure 2-22. Sinkhole in brazing mouth indicating need for more brazing alloy.

Although a tube-within-a-tube connection can be made in any position, the preferred mode is with the brazing mouth facing straight up. In this position the force of gravity will assist in drawing the brazing alloy deep into the joint. A horizontal position is also satisfactory, even though the process is unaided by gravity. Obviously, the least desirable position is with the brazing mouth facing straight down towards the ground. This position allows the force of gravity to work against the opposing forces which attempt to draw the molten brazing alloy upward and into the clearance space of the joint.

It is useful to examine a number of brazed tube-within-a-tube connections to observe the results of the various factors that come into play when making this sort of connection. Figures 2-23, -24, -25 show connections which have been cut out of the tubing line and sectioned to show the paths of the brazing alloy in the clearance space.

Figure 2-23 shows a ⅜-inch OD tube inserted within a ½-inch OD tube and brazed with the mouth facing down. Note how the close spacing at the left enabled capillary action to draw the brazing alloy straight up to the top against gravity. Compare this action with that on the right side where capillary flow had only a minimal effect due to the wide gap. The brazing alloy which did penetrate in this sector was drawn upward by braze spreading, braze shaping and some slight degree of capillary action. In fact, more brazing alloy should properly have been pushed upward into the gap. Although the joint may appear weak upon first inspection, it will hold up well in service. Mechanical strength is provided by the capillary braze on the left while the fillet braze on the right is sufficient enough to seal the remainder of the opening. The additional benefit of a good backup fillet is also illustrated here in Figure 2-23.

Figure 2-24 shows a ½-inch OD tube inserted within a ⅝-inch OD tube and brazed with the mouth facing up. Capillary action, braze spreading, braze shaping, braze strengthening and gravity all combine to fill the clearance space. Note that the relatively wide clearance space at the left is completely

Figure 2-23. Brazed with mouth
facing down.

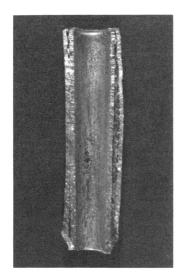 

Figure 2-24. Brazed with mouth
facing up.

Figure 2-25. Brazed with mouth in horizontal position.

filled with alloy. Voids such as those apparent on the right side of the joint depicted are commonly found in brazed tube-within-a-tube connections. They result from a number of factors which include: trapped flux, trapped gas and/or dirt. If the amount of trapped contaminants is minimal, the integrity of the joint (i.e., its strength) is not affected appreciably.

Figure 2-25 shows a ⅜-inch OD tube inserted within a ½-inch OD tube and brazed in a horizontal position. As in Figure 2-24, the same brazing factors combine to fill the clearance volume. However, in the horizontal position, the force of gravity is reduced significantly. Nevertheless, notice that the brazing alloy completely fills the clearance space resulting in a strong joint.

It is apparent that in the process of making a tube-within-a-tube connection a substantial amount of brazing filler metal is wasted. Further, the tube-within-a-tube connection offers little or no significant advantage over a reducing coupling. By way of comparison, a ⅜-inch OD line joined to a ½-inch OD line by a reducing coupling requires approximately 1 ½-inches of .050 × ⅛-inch brazing strip while a typical tube-within-a-tube connection requires substantially more—an average of about 5- or 6-inches. In overall appearance, a reducing coupling lends a neater and more professional looking connection.

An effective modification of the common ¼- into ⅜-inch tube-within-a-tube connection can be fabricated by pinching down the end of the ⅜-inch tube around the ¼-inch tube with vise-grip locking pliers as demonstrated in Figure 2-26. The joint is produced by inserting the end of the smaller tube into the orifice of the larger to a depth of approximately ¾-inch. The arc of the larger tube is then pinched down around the smaller tube with pliers. Once complete, release the pliers then flux and braze the connection. The brazed joint using this process is shown in Figure 2-27.

While it is possible to employ this technique in tubing sizes greater than ⅜-inch, it is not recommended because it is difficult to sufficiently pinch down the larger tube sizes for the connection to work effectively.

Figure 2-26. Fabricating a brazing socket by pinching down the larger tube over the smaller tube with vise-grip locking pliers.

The preceding discussion of the tube-within-a-tube connection is not an endorsement of that particular procedure. Rather, it has been examined for two reasons:

- It documents a long standing practice that has become an acceptable part of piping—one not likely to change in the near future.
- It is useful in that it details practical guidelines for producing an improvised connection in which normal clearance spaces are exceeded, facilitating field repairs.

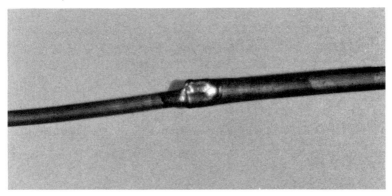

Figure 2-27. Pinched tube brazed connection.

Regardless of the wide spread use of the tube-within-a-tube connection, correct clearance space is a primary, if not crucial, concern in all facets of the brazing process. Increasing the clearance space beyond established specifications results in weak joints. Extreme caution must be exercised when attempting to fabricate a brazed connection with excessive or no clearance space at all. When forced to improvise, it is best to employ only the nonstandard connections that have been proven in the field and have been addressed in the course of this text. However, due to the mercurial nature of field repair, when faced with a new or unique situation which requires fabricating a specialized connection, adhere to the principles presented in the braze spreading, shaping and strengthening discussions as guidelines for improvising the most effective connections.

SURFACE PREPARATION

Mechanical Cleaning

Proper wetting of the base metal is the most essential requirement for effective brazing, as previously detailed in the soldering section. The joint area must be free from oxide films and external contaminants for wetting to occur. As the initial step, tube-ends and fittings should be mechanically cleaned. The cleaning procedures described in connection with soldering are also applicable to brazing.

Mechanical cleaning prior to brazing is particularly important when working with copper tubing which has been in service many years. For example, the typical discharge line, which leads to the condenser, maintains an elevated temperature. The external oxide film that forms on the line tends to thicken with age and/or use. Old on-site tubing may also be coated with materials such as paint, kitchen grease or an oily film which necessitate cleaning before beginning the brazing procedure.

 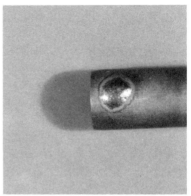

Figure 2-28. Good wetting by brazing alloy.   **Figure 2-29.** Brazing alloy balls up without wetting copper.

It is essential to be aware of how some unusual surface contaminants interfere with the wetting process. In order to observe contaminant interference first hand, clean the ends of two test samples of copper tubing. Next, rub a piece of aluminum foil over the surface of the cleaned end of one sample. Place the other clean sample in a vise and heat until the metal is a dull cherry red. A brazing alloy rod of 15% silver or equivalent readily wets the surface of the copper as shown in Figure 2-28. When the test is repeated on the sample coated with aluminum, the brazing alloy balls up and does not readily wet the copper (Figure 2-29). Without in any way de-emphasizing the need for mechanical cleaning prior to the brazing process, it is safe to say that cleanliness is more critical in soldering copper tubing than in brazing. The primary reason, brazing is carried out at temperatures hot enough to burn-off most of the surface films which normally interfere with the soldering process.

Chemical Cleaning
— Flux

The art of making flux is as old as man's experience in working with metals. Ancient metal craftsmen experimented with many fluxing materials to find those which worked best. Successful flux formulations were zealously guarded secrets, and in time, formed the basis of local industries. Down through the centuries, however, certain chemical compositions became recognized as the preferred flux for a given metal. It was only in the twentieth century that the chemistry of flux began to be thoroughly understood. At the same time, the range of flux was enlarged to include gas atmospheres and vacuum as well as self-fluxing filler metals containing phosphorus and lithium.

The term "flux" is defined as: *a flowing or flow*. Borax or rosin are the most common substances used to help metals fuse together. These substances exhibit the ability to remove surface oxides or tarnish films from the metal surfaces being joined, as well as from the molten brazing alloy, at the moment of joint formation. Flux is not formulated with cleaners to remove foreign contaminants such as grease and dirt. This must be done as a separate cleaning operation prior to brazing. As the definition implies, flux is designed to operate in a narrow critical temperature range to remove those barely visible, thin tarnish film coatings which prevent wetting.

The action of flux in brazing is similar to that in soldering. Although tube-ends and fittings may appear shiny after mechanical cleaning, an invisible oxide film always forms and can subsequently interfere with the brazing operation. This film is formed by oxidation wherein the oxygen present in the atmosphere combines with the metal molecules on the surface of the tubing. Since heat has a definite relationship to the rate of oxidation, the process is greatly accelerated when the tube-ends are heated to the correct brazing temperature.

Brazing operations can, however, be carried out in a "reducing atmosphere" which eliminates formation of oxide film. This type of atmosphere can be simulated in a factory under special enclosures where oxygen scavengers—select gases capable of combining with and removing the oxides, prevent oxidation of the metal surfaces. In some instances, factory brazing is performed in a vacuum to completely inhibit oxidation.

When brazing is carried out in the open, as is typically the case in field installation/repair, oxide film formation is inevitable. Before brazing, the oxide film must be removed and cannot be allowed to reform. This is accomplished by the action of a flux. In addition to its principal function of oxide removal, flux must necessarily melt and flow at a temperature below that of the brazing alloy melting point. Flux must also promote wetting and spreading of the brazing alloy, and be capable of being displaced by molten brazing alloy while still maintaining a barrier to protect metal surfaces from oxygenated air.

In addition to promoting oxide removal and brazing alloy wetting, a good flux must also be capable of penetrating the tight clearance spaces of sweat fittings. Correctly formulated flux must also resist a temporary increase in temperature or a slightly extended heating cycle without losing its effectiveness. In both storage and use, flux should resist deterioration and loss of consistency, although some contamination will inevitably occur in use. Consequently, it is advisable not to carry the same container of flux in a service vehicle over an extended period of time because, in the case of paste, it may dehydrate and the contaminants will become concentrated. Since the cost of flux is so nominal, periodically renew the supply in order to assure the best brazing results.

Flux comes in many forms and shapes. Powder, paste and liquid forms are packaged in containers to be applied directly to the joint area. Brazing rods are coated or cored with a flux to facilitate one-step flux/filler metal applications. Flux is also manufactured with a filler of brazing alloy particles in preformed shapes and sheets so they can be inserted within clearance spaces for one-step brazing.

Of the many flux forms available, paste flux, which is brushed on joint surfaces, has gained the widest acceptance in the trades. Paste flux offers many advantages. Because of its consistency it can be applied to joint surfaces in a controlled manner so the correct amount for the application is used. This is an especially important feature in refrigeration piping due to the fact that flux is corrosive and should be kept out of the refrigeration system. Another advantage, paste flux is less likely to be spilled during application. And since it is a paste, its consistency can be restored when necessary by adding a small quantity of water which also makes it economical.

Paste flux for the refrigeration trade is formulated from a combination of chemicals including: boric acid, borates, fluorides, fluoborates, deoxidizers, water and wetting agents. Although there are variations in the performance of flux supplied by different vendors, most are designed to meet flux specification 3A of the American Welding Society. This spec requires a useful temperature range of at least 1,050 to 1,600°F for low temperature silver-containing brazing alloys.

To apply, brush a thin layer of flux around the tube-end and then insert it into the fitting socket until it bottoms-out. Next, rotate the tube within the fitting to spread the flux evenly. Note, to prevent line contamination, flux is not applied to the internal surface of the fitting socket prior to inserting the tube. When the tube-end is inserted into the fitting socket, any flux applied to the internal surface of the fitting will be scraped off and driven inside the line. In this instance, flux that has been driven into the line then enters the system and becomes an undesirable contaminate.

Figure 2-30. Bubbles and white surface patches begin to appear between 212° and 600°F.

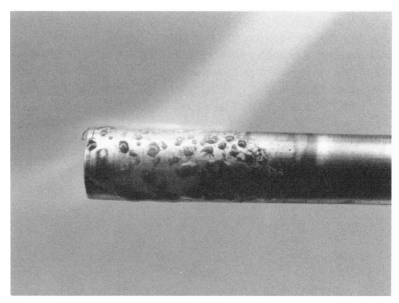

Figure 2-31. At about 800°F flux begins to melt and has a cloudy appearance.

Figure 2-32. At about 1,100°F flux is fully active.

Once properly assembled, the tube-socket connection is then heated to brazing temperature. During the process of heating the assembly, the flux undergoes various changes which serves as a temperature indicator and guide for timing the application of the brazing alloy. Between 212 and 600°F any water present is driven off, and bubbles begin to appear along with white surface patches. At approximately 800°F the flux begins to melt and has a cloudy appearance. At about 1,100°F the flux takes on a clear glassy appearance and is active and ready for the brazing operation. Figures 2-30, -31 and -32 show the three important stages in flux activation. Figure 2-33 is a chart illustrating the behavior of a flux during the brazing cycle.

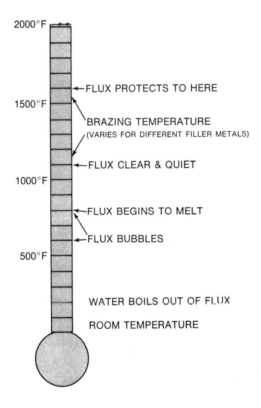

Figure 2-33. Chart showing flux behavior during brazing cycle. (Courtesy, Copper Development Association).

The term "flux" is generally interpreted as describing a special chemical formulation applied as a first step in a two-step brazing operation. However, other materials, such as phosphorus and hydrogen, can also act as a flux. Brazing alloys containing phosphorus are self-fluxing and can be used in the one-step brazing of copper. In fact, phosphorus is the most important fluxing agent in copper tube brazing. Since phosphorus is added to brazing alloys for reasons other than fluxing, its connection with brazing filler metals should be touched upon.

In the reducing flame of a brazing torch, hot hydrogen gas tends to combine with some of the oxide film on copper to effect a partial fluxing action. This feature is addressed in a latter section dealing with brazing torches.

To complete the discussion of flux, mention must be made of flux-coated brazing rods for one-step brazing. In refrigeration piping applications, flux-coated rods may be advantageous for brazing because the rods reduce the chances of flux entering the system. The correct ratio of flux to filler metal in the rod is taken into account during the manufacturing process thereby providing only the amount of flux required for the job. Using this type of rod prevents excessive use of flux which can easily occur when manually brushing paste flux on surfaces to be brazed. And when a number of connections must be made, this one-step application can save a considerable amount of time.

Although flux-coated rods are common in high temperature brazing and welding, they have found little application in low temperature brazing of copper and copper alloy tubing products using silver-containing filler metals. There are two main reasons for this:

- There is some concern that flux applied at the joint mouth will not fully penetrate the clearance space of standard sweat fittings.
- The use of self-fluxing copper-phosphorus brazing alloy greatly reduces the need for flux in brazing copper.

Of course, flux is still required on brass and in brazing ferrous metals with filler metals that do not contain phosphorus.

Flux-coated brazing rods are available for low temperature brazing. Brazing rods of various alloys have been coated with tough, matching flux formulations which do not chip or crumble off the rod when handled. An interesting feature is that the flux coating on these rods is color coded by some manufacturers so the alloy composition can be readily identified. Limited use for these low temperature rods has been found in the typical factory setting, particularly in brazing copper to steel tubing joints. However, the verdict is still out as to whether flux-coated rods will find a more generalized use in field repair and installation work.

BRAZING FILLER METALS

Brazing filler metal is an alloy which characteristically has a melting temperature above 840°F but below that of the metal parts which are to be joined. Alloys must exhibit the property to wet base metals and to form strong, leak-tight joints upon cooling. Further, filler metal must be both chemically and metallurgically compatible with the base metals to prevent corrosion. It also must have the requisite ductility to resist the effects of vibration as well as expansion and contraction caused by temperature changes.

There is no single universal brazing alloy which is compatible for use on all metals. Different base metals, structural joints and heating techniques may all require brazing alloys that demonstrate special properties to suite the application. Consequently, a wide variety of brazing alloys are necessarily manufactured. These are furnished in powder, wire, foil or rod form depending on the type of application.

A successful brazing alloy results from painstaking research which is substantiated by extensive laboratory and field testing. For example, in formulating a brazing alloy for copper the metallurgist starts with a copper base metal which is alloyed with various metals to impart certain properties. The melting point is then lowered by alloying the base metal

118

with either silver, tin, phosphorus, zinc, antimony or a combination of these elements. The ductility is improved by supplementing the base with either silver, tin, cadmium, nickel or a combination of these elements. Certain flow and corrosion resisting properties can also be imparted by the combination of the alloying metals.

Formulating an alloy is more involved than merely adding metals to the base metal in order to achieve a desirable property. Critical interrelationships may exist among the constituents of an alloy. By arbitrarily varying the proportions of certain alloy metals, joint strength may catastrophically be reduced. Optimum results are only obtained when calculated proportions of the supplemental metals are added.

The metallurgist conducts research by running a gamut of tests on the same combination of alloying elements. In each test, the proportions of one or more of the constituent metals is slightly modified. A test similar to the one shown in Figure 2-34 is conducted for each alloy combination. A copper plate is heated by a torch and brought up to brazing temperature. The sample brazing rod is worked back and forth across the plate while its wetting and flow characteristics are observed. The melting temperature (liquidus) and the

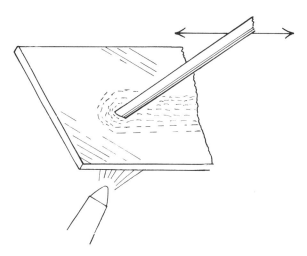

Figure 2-34. Testing a brazing alloy sample.

solidifying temperature (solidus) are recorded. Select joints are then brazed with the sample alloy and subjected to tensile and bending stresses to determine the strength of the alloy. The data from this battery of tests is finally tabulated in chart and graph form and studied to determine whether any trends or unusual metallurgical properties are apparent.

An important breakthrough in brazing alloy technology was made in the early 1920's when Jesse L. Jones of the Westinghouse Electric And Manufacturing Company of Pennsylvania formulated a copper based brazing alloy containing 6 to 10% phosphorus. As the refrigeration industry began to develop in the 1930's and 1940's, phosphorus-copper became an important brazing alloy in both manufacturing and field applications.

Prior to the discovery of the phosphorus-copper brazing alloy copper rotor bars in electric motor manufacturing were brazed with a copper-zinc alloy. The relatively high melting point of the copper-zinc alloy (1,832°F) had the tendency to warp copper rotors and promote hydrogen embrittlement after long periods of heating. This alloy also exhibited a problem when some of the zinc volatilized and formed a zinc oxide coating on the copper that inhibited wetting action.

The phosphorus-copper brazing alloy is a significant improvement over the old copper-zinc alloy. Its approximate 1,310°F melting temperature reduced the brazing temperature by some 500°F which eliminated warping problems and greatly reduced fuel consumption and consequent heating costs. Jones' new alloy also improved flow characteristics and exhibited the trait to readily penetrate capillary spaces which resulted in strong joint formation.

A remarkable property of phosphorus-copper brazing alloy is the self-fluxing characteristic it exhibits when used on copper. At brazing temperature, the phosphorus in the alloy reacts with the copper oxide tarnish reducing it to copper metal and a phosphorus oxide. Eliminating the necessity of using a separate corrosive chemical flux is an especially important advantage when working with electrical conductors.

The properties of a phosphorus-copper brazing alloy

which make it so attractive for electrical work also apply to refrigeration brazing. Because phosphorus-copper alloy is self-fluxing, the chance of flux entering and contaminating the refrigerant lines during brazing is eliminated. And due to the fact that the phosphorus-copper alloy has a lower brazing temperature, the chance of heat-damage to system components such as compressors, expansion and shut-off valves is reduced. The alloy's good flow characteristics make it ideal for penetrating the capillary space designed into sweat fittings. Finally, it is the most economical brazing alloy available because it does not contain silver.

A number of systems have been devised to classify brazing alloys according to their composition. The American Welding Society (AWS) classification system AWS-A5.8 is presented in Table 4 of the Appendix. Under the copper-phosphorus heading, an alloy consisting of 7 to 7.5% phosphorus and the remainder copper is classified as BCuP-2. Interpreting this designation, the B represents brazing, Cu the chemical abbreviation for copper, P the chemical abbreviation for phosphorus, and the 2 a code that indicates the number of constituents. A maximum of 0.15% impurities, listed under the "other elements" heading, is permitted.

Alloy BCuP-2 can be considered an eutectic alloy (i.e., an alloy having a sharply defined melting point). It has a solidus of 1,319°F and a liquidus of 1,460°F, providing a narrow melting range of approximately 150°, which results in a free flowing alloy capable of penetrating tight clearances. BCuP-2 is unsuitable for fillet brazing or filling wide gaps in a clearance space and should only be used on well-fitting joints. Joints made with BCuP-2 can resist moderate vibration.

Use of flux is essential when brazing brass or bronze. And although flux is not required when using BCuP-2 alloy on clean copper, its use will improve joint quality.

Copper-phosphorus alloys are only formulated for use on copper and copper alloy base metals. They should not be used on ferrous metals because brittle phosphide compounds form which tend to weaken joints. This claim can be demonstrated by performing a simple test. Braze two butt-welded joints of ⅜-inch steel rod—one with a copper-phosphorus alloy and another with a phosphorus-free alloy such as 45% silver. Place the ends of both samples in a vise and tap the opposite ends with a hammer. Note in Figure 2-35 that the copper-phosphorus joint snaps with a clean break illustrating its brittle nature while the 44% silver alloy joint resists considerable hammering as demonstrated in Figure 2-36.

Figure 2-35. Failure of steel butt joint brazed with phosphorus-bearing alloy.

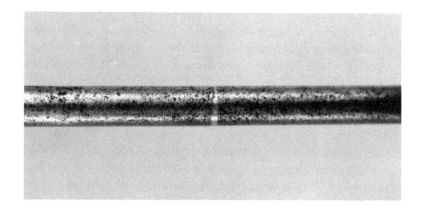

Figure 2-36. Steel butt joint brazed with phosphorus-free silver alloy resists failure.

An important rule to remember when brazing is never use a phosphorus-containing brazing alloy on iron or steel.

The American Welding Society lists another copper-phosphorus alloy BCuP-1 which contains 4.8 to 5.2% phosphorus and the remainder copper. It has a much wider melting range and is used primarily in manufacturing operations with automatic welding and brazing equipment.

Special copper-phosphorus alloys having features tailored to specific needs are also available. For example, the J.W. Harris Company markets STAY-SILV OLP and STAY-SILV OHP which bracket STAY-SILV O with slower and faster flow characteristics respectively.

Figure 2-37 lists the principal suppliers and the trade names of their copper-phosphorus alloy (BCuP-2) which is commonly used in field repair and installation of refrigeration equipment.

AWS CLASSIFICATION	WESTINGHOUSE	HANDY & HARMAN	J.W. HARRIS	ENGELHARD
BCuP-2	PHOS-COPPER	FOS-FLO 7	STAY-SILV 0	SILVALOY 0

Figure 2-37. Principle suppliers of BCuP-2, copper-phosphorus brazing alloys.

Another significant breakthrough in copper brazing technology took place in the early 1930's when Robert H. Leach of Handy & Harman added silver to the copper-phosphorus alloy. The addition of silver supplement served to:

- further lower the melting point
- broaden the flow properties to lend the operator more control
- increase the ductility to offer greater resistance to vibration and expansion/contraction caused by temperature changes

Robert H. Leach's original alloy of 15% silver, 5% phosphorus and 80% copper is marketed by Handy & Harman under the trade name of SIL-FOS. This alloy has a solidus of 1,190°F and a liquidus of 1,475°F with a melting range of 285°F. The alloy can be worked below its liquidus temperature, though some metallic crystals result. Where large gaps exist in a joint or where fillet brazing is required, the heat can be controlled to take advantage of the plastic range to shape and mold the filler metal. On the other hand, in a well-fitted sweat joint, the temperature should be rapidly raised to the liquidus point so the alloy flows quickly to penetrate the far reaches of the capillary gap.

Because of the presence of phosphorus in SIL-FOS alloy it is not necessary to use flux on clean copper, although use of a flux will improve joint quality. Flux must be applied when brazing brass or bronze. As with any brazing alloy containing phosphorus, SIL-FOS should not be used to join ferrous metals because of its tendency to form brittle phosphide compounds.

With some fifty years of proven reliability and operator satisfaction, SIL-FOS has become the standard brazing alloy against which other alloys are rated for performance. SIL-FOS is a copyrighted trade name which belongs to Handy & Harman and can only be used by that company to identify their brazing alloy products. Other suppliers manufacture an equivalent brazing alloy which is identified in the trade as 15% silver brazing alloy.

The American Welding Society classifies 15% silver brazing alloy as BCuP-5. Table 4 in the Appendix describes the alloy as comprising 4.8 to 5.2% phosphorus, 14.5 to 15.5% silver and the remainder copper. A maximum of 0.15% impurities is permitted. Any brazing alloy labelled as meeting AWS specification BCuP-5 is placed in the 15% silver category. Figure 2-38 lists the principal suppliers and the trade names of their alloys.

AWS CLASSIFICATION	HANDY & HARMAN	J.W. HARRIS	ENGELHARD
BCuP-5	SIL-FOS	STAY-SILV 15	SILVALOY 15

Figure 2-38. Principle suppliers of BCuP-5 copper-phosphorus-silver brazing alloys.

Special high-silver content copper-phosphorus-silver alloys are also available that feature properties tailored to fit specific applications. Handy & Harman market SIL-FOS 18 which is a eutectic alloy of 18% silver, 7.25% phosphorus and the balance copper. The alloy has a low melting point of 1,190°F and very fast flow characteristics. The J.W. Harris Company markets STAY-SILV 15LP and STAY-SILV 15HP which bracket STAY-SILV 15 with slower and faster flow characteristics respectively. The J.W. Harris Company also produces PHOSON+ which contains 15% silver, 7.3% phosphorus and the remainder copper. PHOSON+ has a narrow melting range of 1,190°F to 1,205°F and very fast flow characteristics. This alloy is designed to repair pinhole leaks and cracks in brazed joints without disturbing neighboring connections.

Brazing alloys of particular interest to the refrigeration trade produced by Handy & Harman, J.W. Harris Company and Engelhard are listed in Table 5 of the Appendix.

The major deterrent to using 15% silver brazing alloy is its relatively high cost due to the silver content. From its initial introduction to the present, brazing alloy manufacturers have sought to formulate alloys that exhibit properties equivalent to 15% silver but with a greatly reduced silver content. With the extensive use of silver in electronic and chemical industries, research to formulate an adequate, low-silver content replacement has intensified in recent years. Silver is rapidly becoming a scarce commercial commodity, and with limited silver reserves, conservation of the metal is essential. This particularly applies to silver brazing alloys where the silver is not recoverable as it is in silver utensils, coins and many chemical processes. In the future it is likely that economic and conservation pressures will have the effect of increasing the use of the reduced-silver brazing alloys now available and of fostering the development of new silverless brazing alloys.

Responding to the demand for reduced silver content brazing alloys, manufacturers have developed a number of low-silver, copper-phosphorus-silver alloys. Engelhard markets an alloy called SILVABRAZE which is a low cost alternative to a 15% silver alloy such as their own SILVALOY 15.

SILVABRAZE consists of 1% silver, 92.5% copper, 6% phosphorus and 0.5% tin. It has a solidus of 1,180°F and a liquidus of 1,460°F with a melting range of 280°. SILVABRAZE has been formulated to mirror the performance of 15% silver alloys. At the low end of the melting range it can be used for fillet brazing. At the high end it can be used for capillary brazing. It finds use in field brazing of copper and copper alloy tubing connections used in refrigeration piping.

Engelhard also produces a number of low-silver, copper-phosphorus-silver alloys. A number of SILVALOY products are listed in the Appendix, Tables 5 and 6. The numeral which occurs after SILVALOY designates the percentage of silver in the alloy. SILVALOY 6 is formulated to closely resemble SILVALOY 15 and typically finds use in the field brazing of copper and copper alloy tubing connections. SILVALOY 5, 2 and 1 are all low-silver alloys formulated with handling properties to fill the gap between SILVALOY 0 and SILVALOY 6.

The J.W. Harris Co. markets a low-silver content alloy called DYNAFLOW which is a low cost alternate to a 15% silver alloy such as their own STAY-SILV 15. DYNAFLOW is an alloy consisting of 6% silver, 6.10% phosphorus and 87.9% copper. It has a solidus of 1,190°F and a liquidus of 1,460°F with a melting range of 270°F. DYNAFLOW has been formulated to mirror the performance of 15% silver alloys. Its flow characteristics and physical properties make it suitable for both fillet and capillary brazing. DYNAFLOW finds extensive use in brazing copper tubing connections in field installations.

The J.W.Harris Company also markets a number of other low-silver alloys; STAY-SILV 6, 5, 2, and BRAYZON listed in Table 7 of the Appendix. The STAY-SILV 6, 5 and 2 alloys are each bracketed with versions having slightly lower phosphorus content (LP) and slightly higher phosphorus content (HP). This selection of low-silver, copper-phosphorus-silver alloys covers a wide range of flow characteristics and brazing temperatures to fill the needs of factory automated brazing as well as hand held torch brazing.

Handy & Harman markets a low-silver content alloy designated SIL-FOS 6M which is a low cost alternate to their own original SIL-FOS brazing alloy. SIL-FOS 6M is an alloy consisting of 6% silver, 6% phosphorus and 88% copper. It has a solidus of 1,190°F and a liquidus of 1,460°F. Its flow characteristics make it suitable for fillet and capillary brazing of copper piping in refrigeration systems. Closely related to SIL-FOS 6M is SIL-FOS 5 which has 1% less silver and a slightly increased liquidus temperature. SIL-FOS 5 can also be used for fillet and capillary brazing of copper piping connections.

As shown in Table 9 of the Appendix, Handy & Harman markets other low-silver alloys: SIL-FOS 6, 5F, 2, 2M and SIL-FOS 1. These low-silver alloys extend the range of operation to cover specialized factory automated brazing as well as hand held torch brazing.

Phosphorus-Free Brazing Alloys

The amiable coexistence between phosphorus and copper does not apply to phosphorus and steel. As previously emphasized, the phosphorus containing brazing alloys conventionally used on copper piping should never be used on ferrous metals because brittle joints result. An inconvenience arises in that a special class of alloys is required for steel brazing connections.

Copper-to-steel and steel-to-steel brazed connections are regularly encountered in the refrigeration trade, and service personnel must be properly versed in how to make these connections. Copper-to-steel brazed connections are typically located at the stub tubes of some fractional horsepower hermetic compressors and at steel compressor service valves in some of the larger tonnage compressors. Copper-to-steel and steel-to-steel brazed connections are also found in steel condensers on refrigerators and some commercial condensing units. Certain steel components such as liquid and suction line shut-off valves, mufflers, filters and dryers require copper-to-steel brazed connections. Figure 2-39 shows a copper tube brazed to a typical steel shut-off valve. For brazing purposes, copper coated steel stub tubes must be

Figure 2-39. Brazed connection in steel shut-off valve.

considered the same as steel. The high temperature in brazing, combined with the wetting action of the brazing alloy, results in alloy penetration of the copper coating.

In addition to steel piping connections, occasions arise in field repair and installation of equipment that require low temperature brazing of steel components. A compressor mounting stud may need to be rebrazed to its base or repositioned to accommodate a different compressor. One of the arms of a fan support bracket may have cracked due to vibration and requires brazing. Figure 2-40 shows a cracked roller bracket from a sliding drawer in a commercial refrigerator which has been repaired using a high-silver, phosphorus-free brazing alloy. The above are merely examples of the many occasions when low temperature brazing can be a valuable aid in service work.

The brazing alloys used for joining steel in the refrigertion trade are the same as those for industry in general. In the early 1930's metallurgists at Handy & Harman who were working with silver brazing alloys for the silversmith trade formulated a family of silver-copper-zinc-cadmium alloys. These alloys were marketed under the trade name of EASY-FLO and gained immediate acceptance in metal working industries.

EASY-FLO 45 is a member of the EASY-FLO family and is an alloy containing 45% silver, 15% copper, 16% zinc and 24% cadmium. It has a solidus of 1,125°F and a liquidus of 1,145°F with a melting range of 20°F. The high-silver content, along with the addition of cadmium and zinc, imparts to this alloy some unique flow and metal wetting properties. Its melting temperature of 1,145°F makes it one of the lowest melting of commercially available silver bearing brazing alloys. A reduction in melting temperature of some 330°F in comparison to 15% silver, for example, serves as an advantage in brazing components which can be damaged by excessive heat.

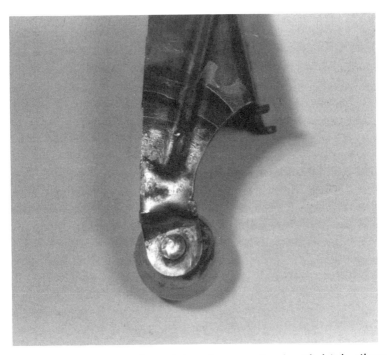

Figure 2-40. Cracked steel bracket repaired with high-silver, phosphorus-free brazing alloy.

The addition of cadmium, along with silver and zinc, permits wetting of steel and most metals with the exception of aluminum and magnesium. The EASY-FLO family of brazing alloys can be used for brazing steel, stainless steel, copper, brass and bronze. EASY-FLO is particularly useful in brazing dissimilar metals such as copper to steel, but a suitable flux is required.

As shown in Appendix Table 5, Handy & Harman market a number of EASY-FLO alloys. These include EASY-FLO 50, 45, 35, 30 and EASY-FLO 3 which are of particular interest to the refrigeration trade. The respective AWS designations are: BAg-1A, BAg-1, BAg-2, BAg-2A and BAg-3.

Engelhard markets a number of high-silver cadmium-bearing alloys which have proven useful in the refrigeration trade. The AWS classification is given in parenthesis after the trade name to identify the constituents of alloy. These include SILVALOY 50 (BAg-1A), 45 (BAg-1), 35 (BAg-2), 30 (BAg-2A), and SILVALOY 25 (BAg-27). SILVALOY 45 contains 5% less silver than SILVALOY 50 but possesses similar handling properties and is recommended by ENGELHARD for general use in refrigeration related brazing. As with all high-silver brazing alloys, a suitable flux is required.

The SILVALOY brazing filler metals are sold in 1, 3 and 5 Troy ounce rolls. The most popular wire diameter sizes are 1/16-, 1/32- and 3/64-inch.

The J.W. Harris Company markets STAY-SILV 50 (BAg-1A), STAY-SILV 45 (BAg-1), STAY-SILV 35 (BAg-2) and STAY-SILV 30 (BAg-2A). As identified by the AWS designations, these alloys generally conform to the standard high-silver cadmium-bearing alloys available to industry. J.W. Harris recommends STAY-SILV 45 and 35 for joining copper-to-steel connections. At present, the company also recommends that cadmium-free brazing alloys be utilized for refrigeration servicing.

Cadmium-Bearing Alloys

Cadmium-bearing brazing alloys are the most versatile and the most widely used of all silver-bearing brazing alloys.

They possess the additional advantage of having proven reliable in over fifty years of service applications. Unfortunately, in recent decades it has been established that cadmium oxide fumes emitted during brazing pose a serious health hazard. Most cadmium poisoning cases have occurred in factory and shop brazing environments where operators were exposed to cadmium oxide fumes over an extended period of time. Occupational safety agencies of the United States and many foreign countries have set a Threshold Limit Value (TLV) for cadmium oxide fume exposure. The TLV in the United States is presently set at O.1 milligram per cubic meter of air for daily eight hour exposures.

Cadmium oxide fumes are odorless and do not elicit any immediate signal as is provided by the pungent odor of escaping ammonia. Cadmium oxide fumes can only be detected by expensive electronic measuring equipment. Consequently, to protect factory and shop workers, brazing areas must be isolated and equipped with suitable ventilating equipment to draw off the cadmium oxide fumes.

Cadmium oxide poisoning is a cumulative process and, much like lead poisoning, requires an overdose or sustained exposure above the threshold limit to cause serious injury.

Cadmium poisoning is primarily associated with brazing in the factory or shop work-place. Since the high cost of installing and maintaining equipment to meet the cadmium TLV can be quite prohibitive, pressure for the development of cadmium-free brazing alloys comes basically from these sources.

Based on practical considerations and without in any way minimizing the cadmium brazing threat, it is safe to point out that the threat of cadmium oxide poisoning is far less in the refrigeration service trade than it might be in a manufacturing environment. Cadmium brazed steel tubing connections occur infrequently in a typical installation. Consequently, brazing a few connections on a daily basis is not likely to produce the cumulative effect that is required to induce cadmium poisoning. In those isolated cases where cadmium

brazing is required simple precautions can ensure safety:
- Keep as far back as possible from the rising fumes.
- Make sure adequate ventilation exists around the brazing zone.
- When possible, use an air-acetylene torch rather than an oxy-acetylene torch since the higher burning temperature of the oxy-acetylene torch generates more cadmium oxide fumes.

The preceding discussion is not to be construed as an endorsement for cadmium-bearing brazing alloys. These observations are based on time-tested fact. Fifty years of use in the repair and installation of refrigeration equipment has proven that cadmium-bearing alloys can be utilized with relative safety. Millions of cadmium-bearing brazed joints can be found in existing refrigeration equipment and serves as testament to that fact.

Inevitably, occasions will arise when a cadmium-bearing brazed joint must be repaired or replaced. Consequently, service personnel must be prepared to work with cadmium-bearing filler metals. As an added precaution, all cadmium-bearing brazing alloys sold today must carry a label that warns end-users to the dangers of cadmium fumes.

In order to meet the demand for a cadmium-free brazing alloy by industrial users, manufacturers undertook an experimental program to replace cadmium with tin and combinations of tin, nickel and manganese. The task was not an easy one because the versatility, wetting and low melting point features of the cadmium alloys cannot be easily duplicated. A large number of replacement alloys were developed to fit the varied needs of industry, and it is likely that continued research will add to the list.

The list of cadmium-free alloys is too long to enumerate here. And many are designed for very specific applications. Figure 2-41, derived from Table 5 in the Appendix, lists a few of the most commonly encountered cadmium-free brazing alloys currently in use.

AWS CLASSIFICATION	HANDY & HARMAN	J.W. HARRIS	ENGELHARD
BAg-5	BRAZE 450	SAFETY-SILV 1370	SILVALOY A-18
BAg-7	BRAZE 560	SAFETY-SILV 1200	SILVALOY 355

Figure 2-41. Cadmium-free, high-silver brazing alloys used in the refrigeration trade.

BRAZING ALLOY SELECTION

Due to personal preferences and the wide variety of alloys available in the marketplace, it is difficult to make specific recommendations when considering alloy selection. Many important factors govern brazing alloy selection, and there is little consensus on which brazing alloy performs best in a particular situation. The diversity of opinion is partially due to the personal prejudices of service personnel and is based, in part, on the inherent variables of the brazing operation. For example, a 15% silver alloy works differently when using an undersized propane-air torch than it does with an oversized oxy-acetylene torch. With the undersized propane-air torch the alloy reaches flow temperature slowly and fills the joint in a sluggish and uneven manner. With the oversized oxy-acetylene torch, on the other hand, the alloy quickly reaches its fluid flow temperature, rapidly spreading and filling the joint.

The important factors entering into brazing alloy selection are:

1. **Service Factor:** The location of the joint in the system is important when selecting a brazing alloy. Two types of service can be distinguished: rough service and average service.

- Rough service is defined as service where vibration is obviously present (i.e., it is seen by the eye and detected by touch). The discharge line to the compressor stub tube connection is a good example of rough service. Other examples are a vibrating suction line/compressor input connection, or a vibrating capillary coil/evaporator inlet connection. Increased temperature and pressure also tend to distinguish rough service, but vibration is the key factor — especially the high frequency vibration that emanates from a compressor.

- Average service is defined as service where no vibration is detected. A typical example is the return bend in a fixed evaporator or condenser. An additional example is the couplings that exist in a properly suspended, remote suction line connected to a compressor. Since the line is remote, any vibrations are damped and do not affect the couplings.

2. **Base Metals:** The nature of the base metal is a key factor in brazing alloy selection.
- Copper-to-copper. The majority of the joints encountered in the refrigeration trade are copper-to-copper, more specifically, ACR copper tubing joined with wrought copper fittings. Silver-phos-copper alloys are preferred in this instance. Their self-fluxing properties and adaptable use with both loose and tight fits clearly lends a distinct advantage. The 15% silver-phos-copper (or its low cost alternates) are preferred for heavy-duty service where the joint clearance is within specifications. For the purposes of field brazing, phos-copper has no advantage over silver-phos-copper alloys other than in its lower cost.

  Although phosphorus-free high-silver alloys, with and without cadmium, can be used on copper-to-copper connections, they are not recommended for field use. The need for flux, tighter clearance space control and higher cost lends them no advantages over the silver-phos-copper alloys currently in use.
- Copper-to-brass. At brazing temperature a portion of the zinc constituent of brass volatilizes in a process known as "zinc fuming." The zinc vapor tends to form a number of compounds, including zinc oxide, which coats the brazing surfaces with a white residue that interferes with the wetting operation. As a result, flux must be used to counteract

zinc fuming when brazing brass. The flux promotes wetting, but it also helps suppress zinc fuming by sealing the brass surfaces.

Field brazing of brass connections requires more skill and know-how than does copper brazing. Due to zinc fuming, as well as a number of other reasons, the use of brass sweat fittings in refrigeration piping has been curtailed and wrought copper fittings are being used instead. However, because of its strength and good corrosion-resistant properties, brass is still the preferred metal for valve bodies and components exposed to corrosive environments.

Since brass is less forgiving than copper, the 15% silver alloys are preferred. These include SIL-FOS by Handy & Harman, SILVALOY 15 by Engelhard, and STAY-SILV 15 by J.W. Harris Company.

The high-silver alloys: BAg-1a, BAg-5 and BAg-7 are also recommended for brass brazing. Note, the term high-silver as used in the context of this book applies to phosphorus-free brazing alloys that contain more than 20% silver. The BAg-1a brazing alloys are marketed by Handy & Harman as EASY-FLO, by J.W. Harris Company as STAY-SILV 50 and by Engelhard as SILVALOY 45. The BAg-5 brazing alloys are marketed by J.W. Harris Company as SAFETY-SILV 1370, by Engelhard as SILVALOY A-18 and by Handy & Harman as BRAZE 450. The BAg-7 brazing alloys are marketed by Engelhard as SILVALOY 335, by J.W. Harris Company as SAFETY-SILV 1200 and by Handy & Harman as BRAZE 560.

When compared with a 15% silver alloy, an important advantage for the high-silver alloys BAg-1a and BAg-7 is their lower melting point. As previously mentioned, most brass brazing in refrigeration piping involves valve bodies that have components which can be damaged by excessive heat. Using a high-silver alloy lowers the brazing temperature by some 300°F and significantly reduces the possibility of heat damage.

Although the 15% silver and high-silver alloys are preferred for brass brazing, for average service conditions the low cost substitutes can be used and may produce satisfactory results. These alternates include DYNAFLOW by J.W. Harris Company, SILVABRAZE by Engelhard and SIL-FOS 6M by Handy & Harman.

Copper-to-steel and steel-to steel. Remember, do not use phosphorus-bearing brazing alloys on steel. High-silver alloys are required when brazing steel.

A wide variety of high-silver alloys are available to fill the needs of industrial steel fabrication. However, the preferred alloys are generally those with a lower melting temperature. Figure 2-42 lists some popular high-silver brazing alloys. The BAg-1a and BAg-2 alloys contain cadmium while the BAg-5 and BAg-7 alloys are cadmium free.

AWS CLASSIFICATION	HANDY & HARMAN	J.W. HARRIS	ENGELHARD
BAg-1a	EASY-FLO	STAY-SILV 50	SILVALOY 45
BAg-2	EASY-FLO 35	STAY-SILV 35	SILVALOY 35
BAg-5	BRAZE 450	SAFETY-SILV 1370	SILVALOY A-18
BAg-7	BRAZE 560	SAFETY-SILV 1200	SILVALOY 355

Figure 2-42. Brazing alloys used on steel.

Service personnel are occasionally called upon to repair ice machines, beverage dispensing equipment and kitchen utensils with components that come in contact with food. Since BAg-7 alloy contains 56% silver and is cadmium-free, the alloy offers a special advantage in that it is approved for use on food processing and kitchen equipment as well as general steel tubing brazing. Even though the 56% silver content increases the cost, its versatility merits a place for BAg-7 alloy on the service truck.

As is the case with brass brazing, flux is required on all steel brazing.

3. **Cost:** As is the case with any material, the cost of the brazing alloy is necessarily a factor in its selection. The cost of a brazing alloy is primary determined by its silver content.

The high-silver alloys are naturally more expensive than the phos-copper alloys or the low cost alternates to the 15% silver alloys. The situation is further complicated by the fluctuating cost of silver and the uncertainty of its future availability.

The question of quantity usage is also a consideration. A pound of brazing alloy can go a long way when working on fractional and low tonnage systems that employ tubing sizes ranging from ¼- to ⅞-inch OD. In a small service and installation operation where brazing alloy consumption is less than 25 pounds per year, it is probably best to utilize 15% silver alloys and their low cost alternates. The greater flexibility exhibited by these alloys in compensating for out-of-round distortions that occur when working with the larger sizes of soft drawn tubing (uncoiled from rolls) lends them a clear advantage over the silverless phos-copper alloys.

A different situation exists, however, in an organization doing large tonnage commercial refrigeration and air conditioning installation work where the liquid and suction lines are comprised of hard drawn tubing in various sizes. The alloy consumption in this instance may run into hundreds of pounds per year making fine tuning of brazing alloy selection economically justifiable.

When it comes to brazing steel connections or components in contact with food there is no choice in the matter. The expensive high-silver alloys must be used.

4. **Fit-ups:** Another important factor in selecting a brazing alloy concerns the condition of the "fit-up." Defined, fit-up is the mechanical and spacing relationships between the parts to be brazed as they relate to the ability of the brazing alloy to make a strong bond. This subject was covered in the section entitled *Capillary Action and Clearance Space,* where it was pointed out that the optimum clearance spacing for copper tube brazing is between .002- to .006-inch. Manufacturers of ACR copper tubing and wrought copper fittings design their products to maintain this clearance space. However, during handling and instal-

lation, copper tubing may become distorted so that the optimum clearance no longer exists. For example, in the act of unrolling a coil of ⅞-inch OD copper tubing the tube might assume an oval shape with a maximum dimension of about .900-inch and a minimum dimension of about .830-inch. When inserted in a standard cup fitting, localized areas exist where the clearance space exceeds .006-inch at the minor axis of the tube and is less than .002-inch at the major axis of the tube. Figure 2-43 shows an out-of-round ⅞-inch OD tube inserted within a standard cup fitting. The gap shown was measured at .022-inch. As mentioned previously, other examples of tubing distortions can also occur.

Figure 2-43. Wide gap in clearance space caused by out-of-round tubing.

When present, distortions impact on capillary brazing to the extent that optimum clearance space no longer exists. A brazing alloy possessing braze shaping properties is then required. For similar reasons, fillet brazing also requires an alloy possessing braze shaping properties.

Most of the brazing alloys used in the refrigeration trade may be employed when optimum clearance space exists. The .002- to .006-inch clearance space can be relied on to draw the molten alloy into the joint by capillary action. The exact brazing alloy selected will then depend on the remaining factors such as base metal, service factor and cost.

5. **Brazing alloy flow characteristics:** Another criterion for selecting brazing alloy is its flow characteristics. Flow characteristics are classified into three categories: slow, medium and fast.

   Slow is defined as the property wherein an alloy moves in a sluggish manner when molten. It allows the alloy to be shaped in order to facilitate filling wide clearance spaces where capillary action alone would not suffice. It also allows an alloy to be used for fillet brazing where the force of capillary action is absent altogether. An example of slow-flow brazing alloys are the 15% silver alloys (AWS designation BCuP-5). These alloys are sold under the trade names of SIL-FOS by Handy & Harman, STAY-SILV 15 by J.W. Harris Company and SILVALOY 15 by Engelhard.

   Fast is used to designate the flow property of an alloy that allows it to move in an extremely fluid manner when molten. A fast alloy has a narrow melting range which allows it to abruptly change from a solid to a liquid state — rapidly spreading-out over the heated base metal. The spreading action of a fast-flow alloy can be likened to the spreading action of penetrating oil when poured on a metal surface.

   A fast-flow alloy requires a good fit-up in order to be properly drawn into the joint by capillary action. It cannot be used for fillet brazing and does not easily lend itself to braze shaping. An example of fast brazing alloys are the phos-copper alloys (AWS designation BCuP-2). These alloys are sold under the trade names of FOS-FLO 7 by Handy & Harmon, SILVALOY 0 by Engelhard and STAY-SILV 0 by J.W. Harris Company.

- **Medium** An alloy with medium-flow characteristics is one that falls between the slow and fast groups previously described. A medium-flow alloy is primarily formulated to give an operator a measure of control over flow rate (not found in the fast-flow alloys) since small gaps are filled most effectively by carefully controlling heat application. Examples of medium-flow brazing alloys are the BCuP-6 alloys sold under the trade names of SIL-FOS 2 by Handy & Harman, STAY-SILV 2 by J.W. Harris Company and SILVALOY 2 by Engelhard.

    The matter of brazing alloy flow characteristics will be addressed in greater detail in connection with liquation and melting range in an ensuing section.

6. **Brazing alloy melting temperature:** Another factor that warrants consideration during brazing alloy selection relates to the alloy melting temperature. Most brazing alloys used with torch brazing melt between 1,100 to 1,500°F. Generally, alloy melting temperature is much more important in automated factory brazing than field torch brazing. Nevertheless, situations arise where a few hundred degrees can make a significant contribution in brazing a satisfactory connection. For example, suppose an expansion valve closely coupled to a multiport distributor requires replacement. Experience proves that heat concentrated in the cramped quarters of the plenum can damage the expansion valve or loosen the tubes brazed to the distributor. Instead of using a 15% silver alloy, a low melting temperature, high-silver alloy may well provide the additional margin of safety necessary to complete the brazed connection without causing heat damage to the valve or distributor.

Examples of Alloy Selection

To conclude this review, additional examples of brazing alloy selection follow:
1. The brazed connection at the discharge stub tube of a compressor is a prime example of rough service. This con-

nection is probably subjected to more stress than any other fitting in a refrigeration system. The high frequency vibrations from the compressor combined with the high discharge temperature stress the connection to its maximum. The 15% silver alloys or their equivalents are preferred here. The silverless phos-copper alloys are not recommended for this application.

2. In the process of replacing a hermetic compressor with a burnt-out motor, installing a suction line filter-drier is standard practice. Filter-driers often have copper coated steel fittings. Consequently, phosphor-bearing brazing alloys are not recommended on copper plated fittings. Rather, the alloys with high-silver content (listed in Figure 2-42) should be used. Flux is also required. Note, in a heavily contaminated system, the filter-drier may have to be changed a number of times, so soldering with 94/6 tin-silver solder may prove to be a more convenient alternative.

3. In an installation employing hard drawn copper tubing where the fit-ups are good and the service factor is average, phos-copper may be used as an alternate to 15% silver. Where the tube-ends have been knocked out-of-round by rough handling or if the connection is subjected to rough service, the 15% silver or alternate alloys are recommended.

4. In an installation employing soft drawn tubing, accurate fit-ups occur infrequently so a slow-flow brazing alloy is recommended for both rough and average service conditions. The 15% silver or its alternate alloys are preferred for all soft drawn copper tube brazing.

5. In brazing copper tubing to brass components such as brass liquid and suction-line shut-off valves on a residential condensing unit the 15% silver or its alternate alloys are preferred. Prior to brazing, the brass valve bodies should be tightly wrapped with a water saturated rag and the flame of the torch should be directed and kept away from the valve body to prevent heat damage. If there is even a remote chance that a valve or its ports may experience

heat damage due to the temperatures required by brazing with a 15% silver alloy, a lower temperature high-silver alloy such as BAg-1a or BAg-7 should be employed as a precaution.

BRAZING ALLOY
WORKING PROPERTIES

Previous chapters made frequent reference to brazing alloy flow properties, braze shaping properties, and melting properties. A better understanding of these factors can serve as an aid to improving brazing techniques. The subject of liquation — the separation of a brazing alloy into liquid and solid components, also needs to be addressed.

A brief review of the constitutional diagram presented in Part I — Soldering, might prove helpful before proceeding. The common soldering alloys for joining copper tubing have related flow and melting properties but the effects are much less pronounced than in brazing alloys. For example, the 15% silver brazing alloy has a melting range of 285°F and if heated too slowly liquation occurs. A 50/50 tin-lead solder has a melting range of only 60°F and liquation is no problem.

Since most refrigeration brazing involves the joining of copper or copper alloy base metals, it is logical to assume that copper would also be part of a brazing alloy. Pure copper has a melting point of 1,981°F which is too high to enable use by itself. It must be alloyed with one or more other metals to lower the melting point and impart other desirable brazing properties.

When copper is alloyed with other metals its melting properties become very complicated. Instead of having a single sharply defined melting point, the alloys exhibit a melting range which can vary from a few degrees to hundreds of degrees Fahrenheit depending upon the nature and the amounts of the alloying elements. In those alloys with wide melting ranges, solid and liquid phases coexist which gives the melt a plastic consistency. This plastic consistency is variously referred to in manufacturers' literature as a *pasty, mushy, tacky* or *semi-solid* state. By adjusting the ratio between the solid and liquid parts the fluidity of the brazing alloy can be controlled to an advantage. This adjustment is achieved by controlling the brazing temperature. A constitutional diagram can help explain.

Most commonly used brazing alloys have two or more metals in addition to the copper base. The constitutional diagrams of three and four metal alloys are quite complicated and difficult to follow. The simplest brazing alloy consists of an alloy of two metals. Copper alloyed with silver is generally selected for demonstration. Figure 2-44 shows the constitutional or phase diagram for the family of silver-copper alloys. When silver is alloyed with copper, the melting point is lowered and the flow characteristics are altered. A picture of what occurs as the alloy melts and solidifies can be seen on the constitutional diagram.

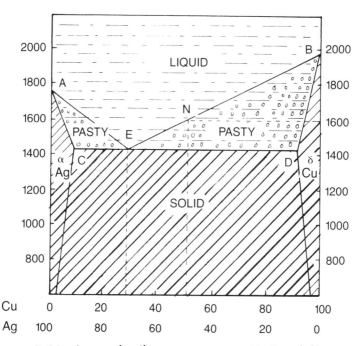

Figure 2-44. A sample silver-copper constitutional diagram.

The vertical lines represent the percentages of silver and copper constituents. The extreme right hand vertical line represents 100% copper and 0% silver while the extreme left hand line represents 0% copper and 100% silver. The whole range of silver-copper compositions are shown between the two extremes. The horizontal axis lines represent temperature as indicated in degrees Fahrenheit on the sides of the diagram.

The diagram is created by taking every silver-copper combination and subjecting it to a controlled heat test to determine the temperature at which the alloy begins to melt and the temperature at which it exists in a completely molten state. The values are plotted and yield a number of curves defining the melting properties of silver-copper alloys.

The curve A-C-E-D-B defines the solidus, which is the highest temperature at which an alloy remains solid. The curve A-E-B defines the liquidus or the lowest temperature at which the alloy is completely liquid (molten). Points A and B are common to both curves. Point A represents the melting point of 100% silver at 1,761°F while point B represents the melting point of 100% copper at 1,981°F. The solidus and liquidus lines divide the diagram into three distinct zones representing the three different states of the alloy. The area below the solidus curve indicated by the 45° hatching represents the solid state. The area above the liquidus curve indicated by the horizontal dashed line represents the liquid state. The area between the solidus and liquidus curves indicated by the dashed lines and circles represents the "pasty state" where liquid and solid particles coexist.

Point E on the diagram is situated at a minimum of the liquidus curve where it touches the solidus curve. This point locates the silver-copper mixture having a precise melting point and is termed a eutectic. From the diagram it can be seen that the silver-copper eutectic has a composition of 72% silver and 28% copper and melts at 1,435°F. This eutectic alloy has the AWS designation of BAg-8 and exhibits the characteristics of a pure metal with sharply defined melting and flow points.

Non-eutectic alloys, however, display a melting range

rather than specific melting temperatures. When heated, non-eutectics begin to melt at one temperature and do not become completely molten until the top of the temperature range is attained. For example in Figure 2-44, an alloy of 50% copper and 50% silver is charted on the diagram as the dotted line in the center of the chart under point N. As the alloy is heated, the temperature increases until the solidus line C-D is reached. The alloy begins to melt at the eutectic temperature of 1,435°F and remains at that temperature until all the eutectic alloy in the system melts out of solution. At this point the eutectic solution is mixed with copper crystals and thickens slightly. As the temperature rises above the eutectic temperature, copper dissolves in the eutectic and the temperature continues to rise along the dotted line until it reaches point N on the liquidus curve. Point N indicates a temperature of approximately 1,600°F at which point the 50/50 silver-copper alloy is completely molten.

When moving from the solidus to the liquidus, the 50/50 silver-copper alloy takes on a pasty consistency as it is heated and passes through the melting range. The actual melting temperature range of the alloy is calculated by subtracting the lowest point of the melting range (1,475°F) from the highest point (1,600°F) which yields 165°F. At the lower end of the melting range the alloy has more body as indicated by the density of the circles which represent solid particles suspended in the fluid eutectic. At this stage the alloy is best suited for braze shaping and clearance gap filling. At the highest end of the melting range, when molten, the alloy has less body. This stage is represented on the chart in Figure 2-44 by having fewer solid particles (circles) suspended in the eutectic. In this condition the alloy is more fluid and can flow into tight clearance spaces.

Since liquation does not occur in eutectic alloys, a eutectic alloy, or one having a relatively narrow melting range, offers many advantages in brazing. One, for example, is that the lower melting temperature and narrow melting range reduce the possibility of heat damage to surrounding components. In addition, eutectics offer fast-flow characteristics that promote rapid spreading and deep penetration between the surfaces to be joined. However, eutectic alloys require a carefully controlled clearance space which cannot always be obtained, especially under field service conditions. As a result, the important alloys such as 15% silver (or its equivalent) which must be used are not eutectic and have a considerable melting range.

The properties of the silver-copper alloys which have been addressed also apply to the basic aspects of phos-copper and silver-phos-copper alloys. The common 15% silver alloy has a solidus of 1,190°F, a liquidus of 1,475°F, and a melting range of 285°F. At the low end of the brazing range the alloy exhibits a slow spreading characteristic and can be shaped easily. Figure 2-45 shows a joint brazed with 15% silver at the low end of the melting range. The copper is heated to approximately 1,300°F at which point it takes on a dull red coloration. Note the extensive fillet around the joint mouth. At the high end of the brazing range the alloy exhibits a faster spreading action and, when heated to that point, flows more readily. Figure 2-46 shows a similar joint brazed at the high end of the melting range. The copper is heated to 1,500°F which produces a pronounced red glow. Note there is no fillet around the brazing mouth. The absence of a filet serves as evidence of the highly fluid nature of the molten alloy.

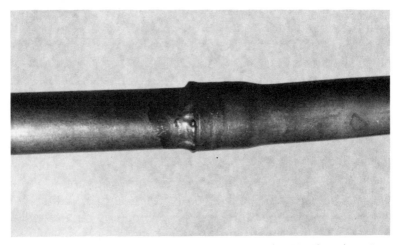

Figure 2-45. Joint brazed in *pasty* stage of 15% silver brazing alloy.

Many of the phos-copper and silver-phos-copper alloys can be worked below their liquidus temperatures. The minute size of the metallic crystals enables the fluid eutectic to entrain the crystals as it is drawn into the capillary space. The ability of the phos-copper and silver-phos-copper alloys to flow well below their liquidus temperatures is an important advantage not found in most other brazing alloys.

Liquation

Liquation is the undesirable, fractional separation of a brazing alloy caused by slow heating. The alloy typically has a wide melting range which allows the lower melting liquid phases to sequentially drain away from the main body of the alloy as it is slowly heated. As a result of the draining action, only residues with higher melting temperatures remain. Liquation can be explained by using a setup similar to one used by metallurgists to test a brazing alloy.

Figure 2-47 shows a sample slug of 15% silver brazing alloy placed on a heated copper test plate. A 15% silver brazing alloy consists of 15% silver, 5% phosphorus and 80% copper. It has a solidus of 1,190°F—the lowest temperature at

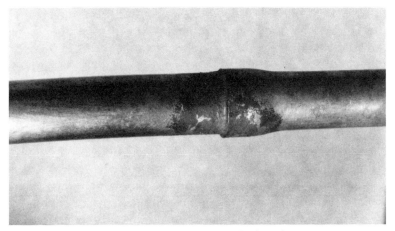

Figure 2-46. Joint brazed at high end of melting range.

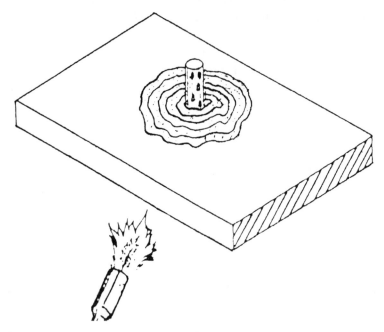

Figure 2-47. Test setup illustrating liquation.

which the alloy begins to melt. Recall from the discussion of Figure 2-44 that the lowest melting point of an alloy occurs at a eutectic which has a definite composition. The lowest melting eutectic composition of the silver-phos-copper series consists of 18% silver, 7.25% phosphorus and 74.75% copper which melts at 1,190°F. As the copper plate is heated and reaches a temperature of about 1,190°F the silver-phos-copper eutectic alloy begins to form and drains away from the main body of the alloy. The eutectic alloy exhibits a very fast flow characteristic and spreads out over the surface of the copper plate. As it drains away from the main body of 15% silver alloy, the eutectic robs the alloy of much of its silver and phosphorus—only a rich copper residue is left behind. The residue has a much higher melting point than the other constituents of the alloy, so the test plate must be heated above 1,700°F to completely melt the remaining alloy.

Figure 2-48 shows a joint brazed with 15% silver where the effects of liquation are visible. The joint was heated too slowly. Consequently, the eutectic drained away from the alloy. The unmelted alloy lumps (copper rich residue called a "skull") at the mouth of the joint are evidence of the slow heating. In the event liquation occurs when brazing a connection and skull is formed, do not attempt to melt the skull by direct application of heat. Due to its high melting temperature, the recommended procedure is to dissolve skull with fresh, molten brazing alloy.

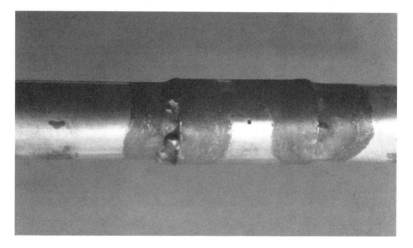

Figure 2-48. Slowly heated brazed joint showing effects of liquation.

To prevent liquation problems, heat the joint to the proper brazing temperature before applying any alloy. Heat the joint until it is a dull cherry red, and simultaneously preheat the brazing rod with the outer most fringes of the flame. Draw the flame slightly away from the brazing mouth (towards the base of the brazing cup) and wipe the brazing rod around an arc of the brazing mouth. If the joint is heated sufficiently and the process is executed correctly, the alloy will melt and be drawn into the clearance space without any alloy separation (liquation). When the torch is finally withdrawn from the joint, it cools immediately. In fact, the normal cooling rate is so rapid that the alloy freezes in place without any phase separation.

Another way of examining liquation is by means of a melting curve. Figure 2-49 shows a simplified melting curve for a sample of 15% silver brazing alloy (BCuP-5) heated on a test plate as illustrated in Figure 2-47. As the alloy is heated, its temperature increases until it reaches a plateau at approximately 1,190°F and begins to melt. Initially, a eutectic alloy of 18% silver, 7.25% phosphorus and 74.74% copper melts. The temperature remains at 1,190°F until all the eutectic alloy drains away from the sample. The remaining alloy, depleted of its silver and phosphorus components, has a sharply increased melting point which approaches that of pure copper — 1,981°F.

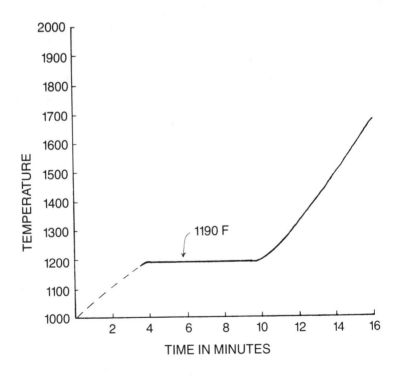

Figure 2-49. Simplified melting curve for 15% silver brazing alloy.

In a practical situation using 15% silver alloy, the eutectic alloy (formed initially) flows into the clearance space. Despite liquation, if the fit-up is good, a satisfactory joint is obtained—even if a skull is apparent on the finished job. In this instance, the 15% silver alloy is converted into a fast flowing eutectic alloy by the liquation process and is used to an advantage. However, if the clearance space on a joint is too great the eutectic alloy may not completely fill the gap, and a defective joint can result.

HEATING and TORCH TECHNIQUES

A brief review of the *Equipment And Techniques* section might prove useful in lending the reader background information on the various torches and fuels employed in manual brazing which will be referred to during the following discourse.

Flame Chemistry

There is little doubt that heat control through torch manipulation is the primary key to successful brazing. In addition, the chemistry of the flame is also of some importance. In most instances, a reducing flame is preferred for brazing. In a reducing flame an excess amount of fuel is supplied so carbon monoxide and hydrogen are generated in the outer flame envelope. These hot products of combustion then combine with the copper oxide film. This "reduction" (combination process) removes oxygen from the copper oxide film and, consequently, yields pure copper.

The ability of the brazing flame to assist in oxide removal is of great importance when working with copper tubing. Recall when exposed to air at room temperature, mechanically cleaned copper surfaces instantaneously form thin oxide films which interfere with wetting. When heated above 600°F copper rapidly oxidizes and forms a flaky black coating of copper oxide. The oxide layers that form interfere with the brazing operation and must be removed. A reducing flame inhibits the formation of copper oxides in the brazing zone and is of particular importance when using silver-phos-copper and phos-copper self-fluxing alloys.

A reducing flame is produced when a slight excess of fuel gas is supplied to the torch. When observing the flame of a torch, the excess fuel tends to form a visible feathery edge around the blue inner cone. This edge supplies the reducing agents to the larger outer cone which, in turn, is the main source of heat for the brazing operation.

Nearly all standard air-acetylene or air-fuel gas torches used in the field produce a reducing flame. The size and position of the torch orifices (air and fuel supply ports) are designed to supply a constant air/fuel mixture that automatically produces a reducing flame. Even though the size of the flame can be controlled, the relative proportions of fuel and air feed to the torch are fixed and cannot be modified.

A typical reducing flame from a number 3 air-acetylene torch is illustrated in Figure 2-50. The blue inner cone is formed by the primary combustion of air and acetylene. The light blue feather surrounding the inner cone is formed by the excess acetylene. The light orange outer zone is where

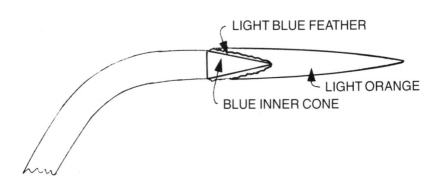

Figure 2-50. Reducing flame from an air-acetylene torch.

the unburnt excess portion of fuel fans out in search of oxygen. The outer most zone is an oxygen scavenging zone and is where copper oxide reduction takes place. During the reduction process, oxygen is liberated from the copper oxide and combines with the excess acetylene. The oxygen/acetylene (air/fuel) combination is then consumed in the combustion process.

Other air/fuel torches generate a similar three zone pattern. The use of larger tips and various fuels result in zone size and color differences which may vary slightly from the air-acetylene pattern of Figure 2-50. However, tips that employ a swirling combustion chamber produce a reducing flames with modified flame patterns. The inner cone and light blue feather seen in the air-acetylene pattern become mixed in a rotary pattern and are not clearly distinguishable. Also, the outer envelope takes on a pale blue cast with a light orange center (Figure 2-51).

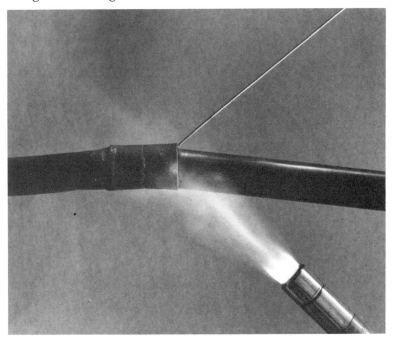

Figure 2-51. Reducing flame formed by swirling combustion chamber.

REDUCING FLAME

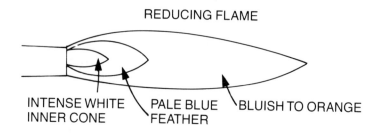

INTENSE WHITE
INNER CONE

PALE BLUE
FEATHER

BLUISH TO ORANGE

NEUTRAL FLAME

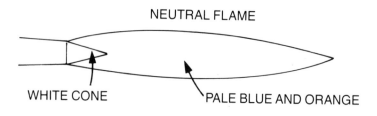

WHITE CONE

PALE BLUE AND ORANGE

OXIDIZING FLAME

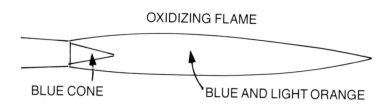

BLUE CONE

BLUE AND LIGHT ORANGE

Figure 2-52. The three flames of an oxy-acetylene torch.

The situation is a little more complicated when considering an oxy-acetylene or an oxy-fuel torch. Because the air/fuel ratio can be adjusted, three different types of flames can result: reducing, neutral and oxidizing. These three flames are illustrated in Figure 2-52.

- The reducing flame is formed by supplying excess fuel which forms a separate feather around the inner cone. The flame pattern here is similar to that of the air-acetylene reducing flame except that the temperature is higher because pure oxygen is utilized rather than ordinary atmospheric air.

- In a neutral flame equal parts of fuel and oxygen are supplied to achieve complete combustion. The resultant products of combustion remain somewhat chemically active. The neutral flame lacks the feather surrounding the inner cone; and the outer envelope is not as wide as that of the reducing flame.

- In an oxidizing flame, excess oxygen is supplied to the fuel to promote rapid and hot combustion. Some of the hot products of combustion act as strong oxidizing agents and can chemically combine with the filler and base metals. The oxidizing flame has a narrow blue or purple inner cone with a long narrow outer envelope. The oxidizing flame is also the noisiest of the three flames, conversely, the reducing flame is the quietest.

The examination of these three oxy-fuel flame patterns were based primarily on the flames of a common oxy-acetylene torch. Other fuel gases combined with oxygen may yield slightly different shapes and colors, but the basic patterns are similar.

There is good reason for not using an oxidizing flame. Figure 2-53 shows a reducing flame directed against the end portion of a piece of copper tubing. Note the white inner cone and surrounding feather. As the temperature rises above 600°F the reducing flame transforms the dark copper oxide film on the tube-end into a bright copper color. When the flame is withdrawn, the oxygen in the surrounding air immediately combines with the copper again and the dark copper oxide layer reforms. Although not shown in Figure 2-53, a back and forth movement of the reducing flame across the hot surface tends to variegate or blanch the metal surface. This is a commonly observed phenomenon when working with copper.

The term blanching is used here to describe the formation of a bright copper surface by the reduction of the copper oxide film when heating copper above 600°F with a reducing flame.

When working with copper, blanching serves to distinguish between soldering and brazing temperatures. The onset of blanching signals that the copper is above 600°F — too hot for ordinary soldering. Blanching is also a sign that the brazing temperature range has been reached. As the temperature of the copper increases, the blanched, shiny copper gradually takes on a dull cherry red coloration and is ready for brazing.

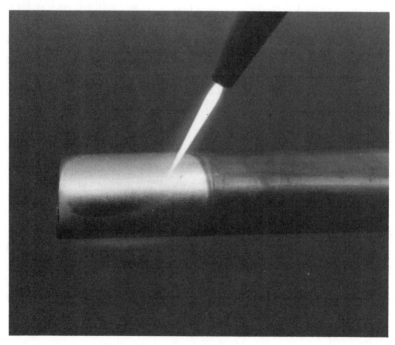

Figure 2-53. Oxy-acetylene reducing flame blanching copper tubing.

Figure 2-54. A 15% silver brazing alloy readily wets blanched copper surface.

Figure 2-54 shows a 15% silver brazing alloy being applied to the blanched surface of Figure 2-53. Note how the brazing alloy spreads and readily wets the copper surface. The reducing flame removes the heavy oxide layer enabling the phosphorus to be more effective in removing the final thin film of oxide coating.

Figure 2-55 shows an oxidizing flame directed against an end portion of a piece of copper tubing. Note the narrow inner cone and narrow outer envelope of the flame. Also note the heavy black copper oxide coating generated by the oxidizing components in the flame. Compare the blackened copper tube of Figure 2-55 with the bright copper tube surface of Figure 2-53 produced by a reducing flame.

Figure 2-56 shows a 15% silver brazing alloy being applied to the surface produced by the oxidizing flame of Figure 2-55. Note how the black copper oxide surface inhibits wetting. Compare this demonstration with that in Figure 2-54 where good wetting is achieved with a reducing flame.

For best results, when using an oxy-acetylene or other oxy-fuel gas, the oxygen/fuel ratio should be adjusted to produce a reducing flame. Remember, a reducing flame can easily be identified by the feathery edge which surrounds the flame's inner cone.

Figure 2-55. Oxy-acetylene oxidizing flame generating black copper oxide coating on copper tubing.

Figure 2-56. Poor wetting action caused by an oxidizing flame and heavy coating of copper oxide.

JOINT HEATING AND FILLER METAL FEEDING

Since the basics of flame chemistry have been addressed, consideration should now be directed to joint heating. Whereas overheating is a problem when soldering; underheating is one of the principle causes of inferior joints in brazing. Accordingly, for the beginner, an oversized torch is recommended for brazing, while an undersized torch is better for soldering.

A typical brazing operation is illustrated in Figures 2-57 and 2-58. A ¾-inch OD bell and spigot connection is shown with an exaggerated clearance space. The bell is formed by a swaging tool which expands the tube-end to provide the proper clearance space. After the tube-ends are cleaned, the spigot is inserted into the bell until it bottoms out. Flux is optional since a phosphorus containing rod is used for the brazing.

The mechanically assembled joint is then heated with a torch and a slow, back-and-forth movement is used across

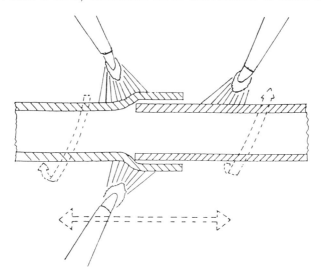

Figure 2-57. Preliminary heating of joint for brazing.

159

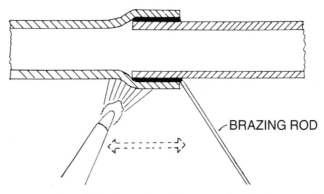

BRAZING ROD

Figure 2-58. Applying brazing filler metal to joint mouth.

and around the bell, as well as out along the spigot about an inch. As the temperature of the copper increases, it starts to blanch, signalling the approach of the brazing temperature range. At this point, the slow sweep of the torch is narrowed to a short back-and-forth movement across the bell. When the copper assumes a dull cherry red coloration, feed the brazing rod into the brazing mouth closest to the flame (Figure 2-58). Provided the temperature of the metal is correct, a section of the brazing rod should melt and be drawn into the clearance space by capillary action. Next, move the torch around the brazing mouth, quickly followed by the brazing rod. In brazing, the rod should be "walked" around the joint mouth to ensure a tight seal. This aspect is where brazing differs from soldering because it is usually sufficient to feed the solder into the solder mouth at only one point. The last step, ring the joint mouth with a small fillet in order to effect a tight seal.

A completed joint is shown in Figure 2-58 where the darkened zone in the clearance space represents molten filler metal penetration.

The proceeding procedure is straightforward and, at least in theory, yields good results in most instances. However, there are a number of variable factors in practical field brazing which may complicate apparently typical situations. These include: joint heat loss, fuel selection and the type and size of the brazing torch employed.

A typical coupling for brazing two sections of ¾-inch OD copper tubing is illustrated in Figure 2-59. In order for silver-phos-copper filler metal to flow into the clearance space by capillary action, the temperature of the coupling must be raised to the 1,300° to 1,500°F range. In theory, only about 50 Btu (British Thermal Units) are required to braze the joint. And by using a number 3 air-acetylene standard tip—adjusted for 6,000 Btu/hr, the job should be done in approximately 30-seconds. Of course in all practicality, a ¾-inch coupling cannot be brazed that rapidly, if at all, with a 6,000 Btu/hr tip. Why? Because the heat is drawn away faster than it can be supplied.

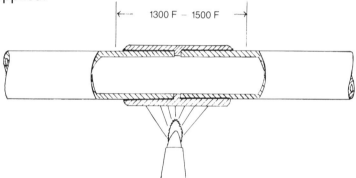

1300 F – 1500 F

Figure 2-59. Standard copper tube brazing coupling.

A brief examination of heat loss is in order. Basically, heat loss occurs in three forms: conduction, radiation and convection.

- <u>Conduction</u> is the principle cause of heat loss in copper tube brazing. Figure 2-60 shows the mechanism of heat loss from a coupling similar to that depicted in Figure 2-59. The arrows running along the copper tubing in both directions from the heat source represent heat conduction. As a metal, copper is an excellent conductor and is capable of conducting a considerable amount of heat away from the joint. Consequently, the temperature of the joint is lowered. The rate of heat conduction depends on the size of the copper tubing and its length or total heat sink. It is easier to braze a coupling or elbow on the end of a copper

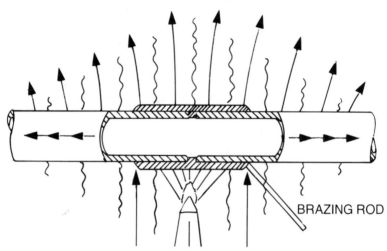

Figure 2-60. Conduction, radiation and convection heat loss during brazing.

tube rather than in the middle where heat conduction readily occurs from both ends of the fitting. Even more heat is conducted away from a joint when located near a compressor body or heat transfer coil.

The effects of heat conduction on brazing can best be observed by brazing a copper tube connection and a similar stainless steel connection side-by-side. The stainless steel connection rapidly heats up to brazing temperature because, as a metal, it is a poor heat conductor. Due to this fact, the loss of heat by conduction is minimal. Moreover, because of its low conductivity, stainless steel tubing is occasionally used to protect brass valves (installed along copper lines) from potential heat damage during brazing. Small sections of stainless steel tubing are often installed between a brass valve body and its connecting copper tubing. These steel sections act as thermal brakes (i.e., road-blocks to the heat being conducted along the copper tubing) which prevent excessive heat build-up from damaging the valve body.

• Radiation is the second type of heat transfer method that accounts for heat loss. In this method, heat energy is transmitted through space as electromagnetic waves that are represented in Figure 2-60 by the wavy lines that emanate from the heated joint in all directions. The most obvious example of radiation takes place on a sunny day. Heat energy from the sun is radiated through space to warm the earth, even though the waves of heat energy must travel many miles to reach the earth's surface. Radiant room heaters work in the same manner to warm living quarters. All hot bodies radiate heat, and the amount of radiation is proportional to the fourth power of the absolute temperature. At elevated temperatures, a sharp increase in radiated energy occurs making heat loss a significant factor in brazing. For the sake of comparison, over 15 times more radiant heat is lost during the brazing process than in soldering.

• Convection is the third type of heat transfer and it occurs by way of a current or flow pattern. This mode of heat transfer is represented in Figure 2-60 by the arrows which point upward to indicate rising air currents. The principle here is the same as that in convection hot air heating. The hot joint and connecting tubing heats the surrounding air which in turn rises and carries heat away from the joint.

Other variable factors that effect brazing relate to the type of fuel gas and torch used in the brazing operation. (For a description of the various fuel gases and torches review the equipment section in Part I — Soldering.)

Of all the fuel gases, acetylene burns with the highest velocity and produces the highest flame temperature. Next, arranged in descending order according to flame temperature, are the enriched LP (liquified petroleum) gases such as MAPP and then propane. Burning the above fuels in oxygen rich environments increases the combustion rate and, as a result, the flame temperature.

While important, flame temperature is not the only factor to be considered when heating a joint prior to brazing. The most important overall consideration is the total amount of useable heat applied to the joint (i.e., target heat). Most of the heat generated by a torch is lost in hot gases which sweep by the tubing and are lost to the surrounding air. The objective in torch design and fuel selection is to concentrate the maximum amount of heat on a relatively small brazing target.

An air-acetylene mixture burned using a standard tip offers the advantage of burning more rapidly, with a hotter flame, and concentrating more heat on its target than can other air/fuel mixtures burned in standard tips. The flame temperature can be increased considerably when burned in an oxygen rich atmosphere (which results in an increase in the rate of combustion and, in turn, a proportional increase in the heat produced). However, the development of the swirl tip some twenty years ago has modified brazing, especially as it applies to the refrigeration trade.

A swirl tip employs a series of curved blades fixed in the tip stem that causes turbulence in the flow of the air/fuel mixture in the form of a swirling motion. The fixed blades of the tip cause the mixture to swirl as it enters an enlarged flame-tube where ignition occurs. This arrangement has two important advantages:

- The swirling turbulence caused by the fins assures thorough mixing of the air and fuel and results in a hotter flame.
- The helical flow path reduces the overall flame length (compared to a straight air-gas feed in a standard tip).

As a result of the swirling tip, heat can be applied to a localized spot where it is needed most. And the short stubby flame, characteristic of the swirl tip, is better suited for tubular brazing. The shortened flame tends to wrap itself around the tube, utilizing a greater percentage of the heat produced, rather than projecting beyond the tube and losing a majority of its heat to the atmosphere.

The swirl tip has made it possible to extend the use of propane and enriched LP gases for all conventional brazing requirements.

The variable factors that have been addressed all interact to control the brazing operation. Figure 2-61 which depicts the heat loss pattern of the ¾-inch coupling illustrated in Figure 2-60 can help explain these interrelationships. The series of temperature graphs presented in Figure 2-61 show the estimated temperature at each point of the coupling for a given fuel and torch.

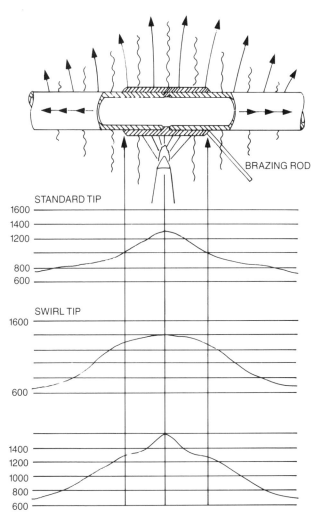

Figure 2-61. Effect of heat loss on joint temperature.

The first graph represents the temperature distribution across the coupling resulting from the use of a number 3 standard air-acetylene torch rated at about 6,000 Btu/hr. The hottest part of the flame is directed at the central region of the coupling to draw the molten brazing alloy into the clearance space. The middle vertical temperature line of the graph coincides with the center line of the coupling. The temperature scale starts at the blanching temperature of copper, approximately 600°F, and increases in 200° increments to 1,600°F. The distance along the coupling, in both directions from the center line, is marked on the horizontal distance scale. Even though air-acetylene burns at a temperature of 4,000°F, the heat loss caused by conduction, radiation and convection is so great that it prevents the coupling from reaching brazing temperature. The hottest part of the coupling is in front of the blue inner cone of the air-acetylene flame. This region may reach a localized temperature of about 1,300°F as indicated on the top graph representing the standard tip. Immediately to the left and right of the peak, the temperature falls, due to heat loss, and is down to 1,000°F at the critical area of the brazing mouth. Note, this temperature is too low to melt the brazing rod.

By moving the flame to a location on the brazing mouth and holding it there for a considerable length of time, it is possible to deposit some lumps of brazing alloy. Even so, true wetting still does not occur. The top graph representing the standard tip illustrates the situation where the torch is too small to overcome heat loss.

The second temperature graph, representing the swirl tip, demonstrates the joint heating mode when a number 3 air-acetylene swirl tip is used instead of a standard tip. A swirl tip burns more fuel and operates at a higher gas pressure than any equivalent sized standard tip. However, the swirl tip lacks the adjustability of a standard tip since a certain minimum fuel pressure is required to maintain the swirling action. In all practicality though, this is only a minor disadvantage.

The advantages of using a swirl tip, as previously mentioned, are the spiralling blunt flame that tends to envelope most of the joint, and the increased flame temperature. The short, broadened flame of the swirl tip heats the entire coupling to a temperature of at least 1,400°F with enough excess heat energy to maintain brazing temperature of the coupling beyond the immediate confines of the brazing mouth. At temperature of 1,400°F, brazing alloy melts readily and is drawn into the clearance space to make an effective joint.

This condition is charted on the swirl tip temperature graph in Figure 2-61 where the peak temperature zone is broadened at 1,400°F level to include most of the length of the coupling. Notice the temperature then begins to drop slightly at the point of the brazing mouth.

The third temperature graph of Figure 2-61 depicts the joint heating method when an oxy-acetylene torch is used to braze the coupling. For a fair comparison, the oxy-acetylene torch selected for this example has a heating range between 6,000 to 10,000 Btu/hr—comparable to the air-acetylene torch used in the previous examples. The oxy-acetylene flame is narrower than the swirl flame but burns at a much higher temperature (6,000°F). The hotter flame compensates somewhat for the deficiency in the flame pattern by applying more target heat to a smaller area.

After the coupling is brought up to near brazing temperature, the flame is then directed at the center of the connection to draw the brazing filler metal inward from the brazing mouth. The hot flame raises the temperature of the central region between the 1,500° to 1,600°F range. The temperature rapidly begins to drop, but the temperature at the brazing mouth maintains at approximately 1,300°F. The increased local temperature at the central portion of the connection tends to duplicate the effects of the short, broadened swirl tip flame. This tendency is displayed in the oxy-acetylene temperature graph where the rather narrow high temperature peak (which dissipates rapidly) is effective in maintaining a brazing temperature of at least 1,300°F at the brazing mouth.

The three temperature graphs shown in Figure 2-61 are a few examples of the many heating situations which can be encountered. The coupling used in the temperature graphs of Figure 2-61 can be considered the equivalent of two cup and spigot connections mounted back to back. In all three examples, the primary consideration is to recognize the need to heat the entire joint to correct brazing temperature. However, most importantly, the brazing mouth must have a lower temperature than the central portion of the connection to promote effective capillary action.

Moreover, when brazing a joint near a large heat sink

such as a compressor or condenser, heat loss through conduction should be taken into account when making the torch tip selection. Table 10 in the Appendix consists of a few charts that recommend tip sizes for various fittings from torch tip manufacturers.

The effect of heat sink conduction on tip size is illustrated using the case of brazing a ¾-inch OD suction line into the suction stub tube of a hermetic compressor using 15% silver brazing alloy. The PREST-O-LITE chart in Table 10 recommends a number 3 SWIRLJET acetylene tip for brazing ¾-inch fittings. A number 3 SWIRLJET tip will braze an isolated ¾-inch OD cup and spigot connection in under 1 ½-minutes. This same torch will braze a ¾-inch OD suction line into a ¾-inch compressor stub tube in about 2 ½-minutes. In the process, the compressor shell in the neighborhood of the stub tube will reach a maximum temperature of 390°F which is too high. This same stub tube can be brazed with a number 5 SWIRLJET tip in about 35-seconds with a maximum shell temperature of only 250°F.

When brazing close to a large heat sink it is good policy to select a larger tip size to complete the brazing operation before any nearby structure, such as a compressor shell, can overheat.

Alloy flow temperature is another factor that deserves careful considered when brazing close to a large heat sink or a component that can be damaged by excessive heat. A difference in the flow point of only 150°F between brazing alloys can mean a sharp reduction in the amount of target heat a joint requires. When a 45% silver brazing alloy with a flow point of 1,145°F is substituted for the 15% silver alloy in the previous example using a number 3 SWIRLJET tip, the ¾-inch OD suction connection is made in less than 2-minutes and the shell reaches a maximum temperature of only about 280°F. A difference of only 155°F in flow points makes a significant difference in both brazing time and in the heat sink temperature increase.

BRAZING FILLET RING

Underheating

The principle cause of poorly brazed tubing connections can largely be attributed to underheating. The proper temperature range from the brazing mouth to the base of the cup and spigot connection must be maintained while brazing, otherwise the joint is weakened. This temperature rise is needed to draw the brazing filler metal into the deepest reaches of the clearance space.

A frequent mistake made when using an oxy-acetylene torch is to concentrate the flame at the joint mouth, heating it until it glows brightly. Then when the brazing rod is applied, it quickly melts around the brazing mouth. Once the brazing mouth has filled, the molten alloy runs over and balls up on the outside of the cup. When the brazing mouth is positioned horizontally or faces downward, alloy tends to ball up on the outside of the tubing.

Figure 2-62 shows a brazed connection at the suction stub tube of a compressor. The brazing mouth was overheated, causing the alloy to ball up on the outside of the cup as shown. When the fitting was disassembled it was found to

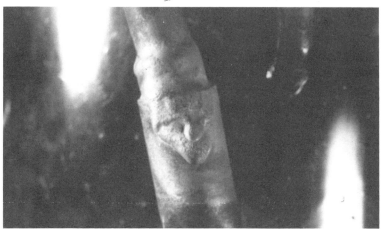

Figure 2-62. Ball up on outside of suction stub tube of a hermetic compressor.

be joined only by a small brazing ring and fillet around the upper portion of the cup. Figure 2-63 was taken by looking into the cup. This photo shows the small brazing ring at the top portion. The lower portion of the cup was not hot enough to draw the brazing alloy to the bottom. And although the brazing mouth was overheated, the deficiency was caused by underheating the lower portion of the cup. Consequently, an adequate amount of molten filler metal was not drawn into the cup.

Figure 2-63. Internal view of a stub tube from Figure 2-62 showing brazing rim around mouth.

Figure 2-64 shows another brazed joint on a compressor where the operator overheated the brazing mouth. This overheating caused the alloy to ball up on the outside of the cup.

Another brazing deficiency caused by underheating is shown in Figure 2-65. A ⅜-inch OD discharge tube was brazed into the discharge stub of the compressor. The heat of an oxy-acetylene torch was concentrated at the brazing mouth. Subsequently, a large fillet built-up around the mouth and extended over onto the tube wall. The connection in Figure 2-65 differs from those in Figures, 2-62 and -64 in that the temperature at the brazing mouth was kept just below the point where the brazing alloy could run down the fitting.

Figure 2-64. Ball up on discharge line connection of a hermetic compressor.

The connection in Figure 2-65 developed a hairline crack around a section of the brazing mouth after about 12 years of service. Upon disassembly, it was discovered that no brazing alloy was present in the lower portion of the cup or spigot.

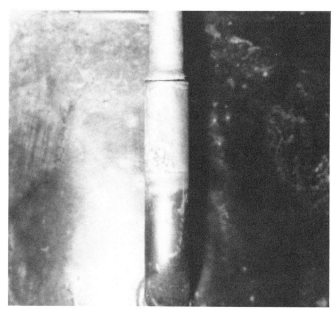

Figure 2-65. Discharge tube brazed to discharge stub tube with fillet only.

This is demonstrated in Figure 2-66 where the spigot was raised vertically, exposing it to view. The joint held for 12 years on a fillet alone. Bear in mind that the discharge connection of a compressor is probably subjected to the most severe service of any connection in refrigeration piping. Consequently, this serves as proof that a brazing fillet does add significant strength to a brazed joint.

Figure 2-66. Discharge tube of Figure 2-65 raised to show absence of brazing alloy on spigot.

The presence of alloy *ball-ups* on the outside of the connection (away from the brazing mouth) serves as an indication of poor heating procedures. Ball-up is a condition which signals the temperature of the base metal was too low for proper wetting action to occur. It is noteworthy to mention that the surface tension of molten brazing alloy is greater than its attraction for a base metal. This fact explains why an alloy draws itself into a ball or round lump when the metal surface it encounters is heated insufficiently to induce proper wetting.

Figure 2-67. Schematic showing of the formation of a ball up.

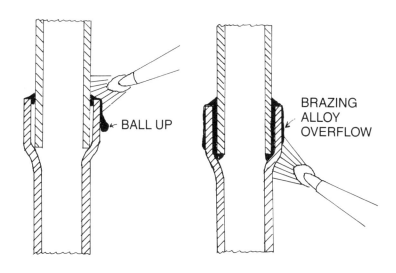

BALL UP

BRAZING
ALLOY
OVERFLOW

Figure 2-68. Brazing alloy has overfilled clearance space and coated outside of fitting.

Figure 2-67 is a schematic representation of a ball-up condition. The hot oxy-acetylene flame is centered at the brazing mouth causing it to overheat and exhibit a bright red glow. When the brazing rod is applied, it melts instantly, streaks around the top portion of the mouth and overflows down the sides of the connection. As soon as the alloy reaches a cooler section of tubing, it stops flowing and pulls itself into a round lump. Note that the filler metal does not penetrate to the base of the cup. The result, a weak joint.

Joint failures can occur at any connection in a refrigeration system but are most frequently encountered at the compressor connections. The main reason—the compressor body serves as a large heat sink drawing heat away by means of conduction. Under these circumstances, concentrating the heat at the base of the cup and spigot instead of at the mouth takes on greater importance. Most of these connections are made using an oxy-acetylene torch on an assembly line in a factory setting. If the brazing operator is not sufficiently trained, poor quality joints can result because of the brief period of time allotted for each brazing operation.

It is important to draw a distinction between molten filler metal which has over filled the clearance space and run down the sides of the connection, and filler metal which has run out of the brazing mouth and balled up on the outside of the connection without filling the clearance space. In the first instance, the brazing filler metal has spread evenly around the outside of the connection by wetting and braze spreading. Although this wastes brazing filler metal and results in a sloppy looking connection, evenly spread external brazing alloy indicates that the joint reached proper brazing temperature and is basically sound. Figure 2-68 is a schematic representation of an overfilled brazed joint. When the brazing alloy balls up or is lumpy on the outside, on the other hand, all indications point to the fact that the brazing alloy did not penetrate to the base of the cup and spigot.

Excessive Clearance

Excessive joint clearance is the second most prominent cause of poorly brazed tubing connections. As mentioned in the *Fit-Ups* section, in the process of uncoiling and stringing rolls of copper tubing, the tubing can become severely distorted out-of-round. Even end portions of straight lengths of hard drawn copper tubing can become distorted by dropping them on a hard surface. Likewise, copper fittings can be distorted by rough handling.

On the popular ¾- and ⅞-inch OD tubing sizes, any out-of-round distortion is easily observed. Because of the large diameters, these two tubing sizes assume a pronounced oval shape after being uncoiled. The actual differences in diameter can easily be measured with a vernier caliper. The out-of-roundness is most apparent when attempting to cut the tube with a tubing cutter. The cutter places two initial nicks across the major oval diameter while skipping over the minor diameter of the oval. Figure 2-69 shows a cutter nick formed on

Figure 2-69. Nick formed by tube cutter on out-of-round tubing.

the outside diameter of a piece of ¾-inch tubing, indicating out-of-roundness.

Brazing can tolerate wider clearance spaces than soldering, but nonetheless, when the gaps are in excess of about .010-inch, even the advantages of better wetting and spreading may fail.

A surprisingly large number of brazed joints on soft drawn tubing have voids in the clearance space where the brazing alloy failed to penetrate. This can be confirmed by unbrazing a number of previously brazed joints on soft drawn tubing. In unbrazing the connection, the spigot should be drawn straight out without any rotation or rocking motion. By rotating or rocking the spigot, the molten brazing alloy will cover the previously unfilled voids and negate the opportunity for visual inspection.

Figure 2-70 shows a ¾-inch disassembled cup and spigot connection. Once separated, a large void on the surface of the spigot is revealed. Another void, not visible in Figure 2-70 is diametrically opposite the one shown. When measured with a precision vernier caliper, the diameter where the voids occurred measure .730-inch. The diameter, at right angles to the void diameter, measured .767-inch. The minimum diameter was .750 − .730 = .020-inch undersized, while the major

Figure 2-70. Disassembled cup and spigot connection revealing large brazing void on spigot surface.

diameter was .767 – .750 = .017-inch oversized. A pair of voids was located inside the cup matching the voids on the spigot. Figure 2-71 is a view looking into the cup of Figure 2-70 showing one of the voids.

The ridge shown on the spigot formed the fillet around the brazing mouth. Note, below the fillet a ³⁄₁₆-inch wide band of brazing alloy extends completely around the girth of the

Figure 2-71. Brazing void on inside of cup matching spigot void.

Figure 2-72. Disassembled coupling showing effects of clearance space on braze quality.

spigot. A matching band of brazing alloy rings the top portion of the cup. Between the voids on the spigot and cup the brazing alloy extends down to the base of the spigot and cup.

The wetting pattern of the brazing alloy on the cup and spigot chronicles the brazing as it took place in the joint depicted in Figure 2-70. The voids on the spigot and cup were obviously caused by excessive clearance due to the spigot being out-of-round. Along the major diameter, where the clearance was tight, the brazing alloy was drawn to the base of the cup. This indicates that the joint was heated properly. The cause of the subsequent voids, however, was not due to underheating but to excessive clearance. Some measure of joint strength was provided by the two brazed sections surrounding the voids. Added strength and a gas tight seal were provided by the fillet and brazing bands that surrounded the brazing mouth.

Another demonstration of the effects of excessive clearance can be seen in Figure 2-72 which shows a ¾-inch brazed coupling that has been separated to reveal the wetting pattern on the spigots. A ¾-inch hard drawn tube with a uniform outside diameter of .750-inch is shown at the left. Note the even braze coating on the spigot and the absence of any voids. A ¾-inch soft drawn tube with a typical oval distortion is shown on the right for comparison. Observe the large void on the spigot. A second matching void exists on the opposite side of the tube on the minimum diameter. Across the major diameter of the tube the brazing alloy penetrated to the base of the spigot. The soft drawn tube was clearly sealed to the coupling by a gas tight brazing fillet and a narrow band that surrounded the coupling mouth.

Careful examination of a number of brazed copper tube connections may reveal the presence of large voids caused by underheating and excessive clearance. Further study of brazed joints that have been in service for over twenty years may indicate that those joints provided with a fillet around the brazing mouth held up well—despite the presence of fairly large voids on the spigots.

In a brazed joint where the clearance space is correct and proper heating techniques are practiced, the clearance space fills with brazing alloy and a fillet is functionally not necessary. However, experience proves that in far too many cases, ideal conditions (i.e., the absence of voids) are rare. Consequently, fillets are necessary and should be added routinely. Remember, the purpose of a fillet is not to provide primary strength but rather to compensate for localized voids by sealing the area around the brazing mouth above the void.

In brazing copper tubing in sizes greater than ⅝-inch OD, the torch and brazing alloy should be walked completely around the fitting, doing one section at a time. No part of the brazing mouth should escape being wiped by molten brazing alloy.

Starting with a cold, clean, mechanically assembled joint, its temperature should be raised to the blanching range with its characteristic copper colored hues. The next step is to concentrate the flame along a section of the fitting, sweeping it back and forth until it assumes a dull cherry red coloration. If flux is used, it should be in the watery stage. Wipe the portion of the brazing mouth in the heated section with brazing alloy while, simultaneously, concentrating more heat at the base of the cup than at the brazing mouth. If an oversized torch is used, do not overheat the fitting so that it glows bright red.

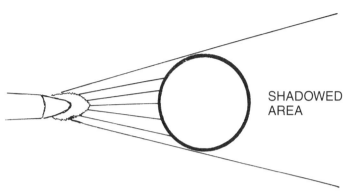

Figure 2-73. Radiant energy shadowing effect on rear side of brazed joint.

After the brazing alloy fills the clearance space of the heated joint, a visible ring forms at the brazing mouth. Once the ring forms, shift the torch to an adjacent section and repeat the operation. Although preferred, but not essential,

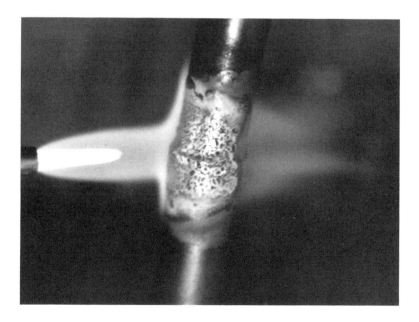

Figure 2-74. Shadowing effects shown during brazing operation.

braze sections sequentially around the joint. The most crucial consideration is that each section overlaps its adjacent sections.

A factor that operates in favor of sectional brazing is the radiant energy produced by the torch. Like light, radiant energy from the flame of a torch travels in straight lines and is reflected or absorbed by a surface it strikes, leaving a shadow behind. Because of the effects of radiant energy, the section of the joint facing the torch gets much hotter than the shadowed section behind it. Radiant energy shadowing is illustrated in Figure 2-73.

Figure 2-74 shows the shadowing effect in an actual brazing operation. The brazed area facing the torch is partially molten as shown by the shiny color. The rear area of the joint has not started to melt as indicated by the darker color. Also, note how the flame itself is divided by the joint. This division leaves a flame shadow behind. Because of shadowing a flame that may normally be adequate cannot place enough heat on a large shadowing surface. And, in a hot flame such as one generated by an oxy-acetylene torch, the side of the joint facing the flame may quickly overheat to a bright cherry red and

Figure 2-75. Brazed joint provided with a fillet.

cause the brazing alloy to scatter while the rear, shadowed side remains below brazing temperature. In these situations, sectional heating is necessary.

The final step in the brazing operation is the application of a fillet. It may be that an adequate fillet has resulted from wiping the brazing rod around the brazing mouth. If not, an external fillet should be supplied for added insurance against leakage. For proper fillet brazing, the temperature of the brazing alloy must be lowered so the alloy exists in the pasty stage. This can either be done by cutting back on the size of the flame or by brushing the flame across the brazing mouth at a controlled rate to dissipate a portion of the excess heat. Figure 2-75 shows a brazed joint provided with a fillet.

On tubing sizes smaller than ⅝-inch OD, the torch flame envelopes the entire joint making sectional heating less important. However, the brazing rod should still be wiped around the brazing mouth and a fillet added.

In some field brazing situations a portion of the joint may not be accessible to a straight brazing rod to allow for continuous wiping around the brazing mouth. For example, a discharge line from a compressor may be blocked on one side by the compressor and a wall of the compressor com-

Figure 2-76. Brazing rod shaped to reach behind obstructed brazed joint.

partment. When confronted with such situations, the brazing rod can be easily shaped by heating with a torch until it softens enough so it can be bent to the desired configuration. Figure 2-76 shows a brazing rod shaped to reach behind an obstructed joint.

ADDENDUM

Disassembly of Brazed Joints

Separating a brazed joint is an easy matter. The difficulties encountered in separating a soldered joint (described in the soldering section) are typically not encountered when dismantling a brazed connection. At brazing temperature, copper becomes quite soft. When a connection at brazing temperature is pulled apart, rather than seizing, the copper flows over resistance points.

Any joint which needs to be disassembled at brazing temperature should be fluxed, even though self-fluxing alloy may have been used when the joint was brazed. Application of flux prior to disassembly prevents heavy oxide coatings from forming on the joint surfaces due to the supplemental heating required.

When rebrazing a connection, the coatings of brazing alloy on the spigot and cup usually make it impossible to mechanically reassemble the joint cold. Brazing alloy cannot be wiped off tubing cleanly as can solder. Consequently, the joint must be assembled hot. To reassemble, butt the spigot up against the cup opening and heat to proper brazing temperature. When the alloy melts, the spigot should easily slide into the cup. Add extra brazing alloy, for good measure, to complete the operation.

Cleaning After Brazing

Copper tubing connections brazed with self-fluxing, phosphorus containing alloys normally require no post-brazing cleaning. The copper oxide formed on the metal surface as the joint cools flakes off eventually leaving a reasonably clean joint surface.

When used, flux should be removed after the brazing operation. The glassy coating that results from using flux can temporarily plug a pinhole leak that may escape initial leak detection, only to show up after the flux coating wears away. Since flux tends to absorb excess impurities and oxides, it turns an unsightly reddish black color which is an easily visible indication that flux is still present on the metal surface.

Most glazed flux can be removed from metal surfaces with abrasive cloth, steel wool or a wire brush. When a glazed lump of flux stubbornly clings to the metal surface, it can usually be scraped or chipped off with a knife or the edge of a file. A well brazed joint should be able to withstand a certain degree of rough handling.

Nitrogen Sweep

When copper is subjected to temperatures above 600°F (as is the case in brazing) and exposed to the open atmosphere, a black, flaky oxide forms on the copper surface. The amount of oxide that forms is dependent upon the temperature, and the length of exposure to temperatures in excess of 600°F . Unlike oxide that forms on external surfaces, which eventually flakes and falls off, oxide that forms on the inside of a tube cannot be removed easily. As a result, the oxide becomes entrained in the system. In large amounts, oxide flakes can not only contaminate a system but, more importantly, excessive build-ups can gradually obstruct or block line and/or valve strainers.

Equipment manufacturers suggest using a nitrogen sweep to prevent the formation of internal copper oxides. In a nitrogen sweep, nitrogen from a cylinder is fed through a regulator, to reduce pressure, and into the piping system while it's being assembled. The nitrogen displaces the air and maintains a slightly positive pressure to prevent re-entry of atmospheric air.

A nitrogen sweep is effective in eliminating internal oxides, but its use involves some heavy equipment and introduces potential safety hazards. A nitrogen tank is bulky and must be handled with care when transported. Accident statistics (especially in the refrigeration service industry) point out that the use of high pressure gases is one of the leading causes of fatal field service accidents. Compressors have been damaged and even destroyed due to insufficient pressure regulation (i.e., high pressurization) or because of a leaking regulator. Most accidents involve inexperienced personnel who do not fully comprehend the hazards involved in the use of high pressure gas. As a standard precaution against such accidents, an experienced service engineer should always be present to verify that the nitrogen sweep system has been installed correctly.

The use of a nitrogen sweep is a matter of practicality. On a large installation, such as in a supermarket where many connections are required, a nitrogen sweep should be included in the installation. The equipment and personnel to transport and set up the sweep are usually available for an installation of this magnitude. On the other hand, for a small service contractor operating out of an automobile or a small truck, whose brazing needs typically involve changing a compressor or repairing a leak, a nitrogen sweep is not essential although it may be desirable. Many years of experience and countless compressor changes have demonstrated that a system can sufficiently tolerate entrainment of the small amounts of copper oxide generated in brazing two or three connections.

Between the two preceding extremes just described, there are installations where a nitrogen sweep may or may not be essential. As rule of thumb, when in doubt, use a nitrogen sweep. If a sweep is not possible, protect the compressor with a suction line filter-drier, and the metering device with a liquid line filter-drier.

In connection with the use of nitrogen to prevent internal copper oxide formation, it is noteworthy to mention that the Mueller Brass Company markets ACR hard drawn tubing (sizes ⅜-inch through 3⅛-inch OD) which is pre-pressurized with nitrogen. Sealed with special reusable caps, these caps can be used to isolate any part of the system during fabrication (Figure 2-77). With practice, this system offers an attractive alternative to nitrogen sweeping and eliminates the need of transporting large cylinders of nitrogen.

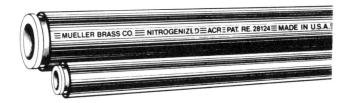

Figure 2-77. ACR copper tubing charged with nitrogen and sealed with reusable plugs (Courtesy, Mueller Brass Co.).

Terminology

The introductory portion of this text addressed terminology associated with the soldering and brazing processes. Within that discourse mention was made that the term "silver soldering" has become obsolete in describing a brazing operation. The use of silver bearing solders such as STAY-BRITE and SILVABRITE has given rise to confusion when the terms "silver soldering" and "silver solder" are applied to brazing.

There have been a number of situations when an engineer, selecting an alloy for joining copper tubing in an installation, has specified "silver solder" for a particular job when the intent was to specify a brazed piping system. In this instance, the contractor who is doing the installation can easily misinterpret the term as meaning silver bearing solder (such as STAY-BRITE or SILVABRITE). Even though in error, by using silver bearing solder the contractor has conceivably met the conditions of the contract.

The terms "silver soldering" and "silver solder" should be abandoned in both soldering/brazing literature and trade terminology. If the presence of silver in the brazing alloy needs to be emphasized, use of the terms "silver braze" or "silver brazing alloy" would prove to be much less confusing.

Brazing Safety

Over the past 50 years, no substantial brazing related safety or health hazard has been discernable. The problems and/or complications associated with brazing which occasionally arise are known entities—for the most part, and can usually be dealt with through the application of common sense and simple precautions.

Adequate ventilation, for instance, is important. At the high temperatures required for brazing, toxic fumes may be generated, particularly when working with brass, bronze or metals coated with zinc or cadmium. Of course, lead fumes generated by any source must avoided. And brazing flux may even generate irritating fumes.

Approach the problem of adequate ventilation on a practical basis. When working out-of-doors where a majority of brazing is done, as a matter of course, simply stand back away from the rising hot gases generated at the joint. Exercising this practice is sufficient in avoiding inhalation of the fumes. When working indoors, consider the number of joints that are to be brazed over a given period of time. For example, if four or less joints need to be brazed throughout the course of a day, which is typical in the majority of field service, simply work at arms length and avoid inhaling the fumes. On the other hand, if the job involves brazing many joints in a confined area, such as a supermarket machinery room, set up a temporary forced ventilation system so the fumes are expelled out-of-doors. In a situation where a joint must be brazed in a tightly confined corner and where forced ventilation is not practical, complete the braze as quickly as possible and leave the area until the fumes have a chance to dissipate.

The special precautions involved in working with cadmium bearing brazing alloys have been addressed in the section concerning brazing filler metals. Review that section when necessary.

Also remember, never braze a system under pressure. At the high temperatures required for brazing, the copper tubing is softened and will not hold the normal pressures of a refrigeration system. In addition, recall that the high temperature tends to break down the oil and refrigerant in the system generating corrosive products.

One other warning in particular bears repeating. Never use oxygen or any fuel gas to pressurize a system for leak detection. Serious, often fatal, explosions can occur. At the same token, it is also wise to never use oxygen or fuel gases as a source of pressure for cleaning or for any other forced-air purpose.

Short of causing explosions, fires have been caused by brazing in close proximity to combustible surfaces. When concentrating on the brazing operation, it is easy to overlook the effect of the flame that sweeps by the brazing site and comes in contact with nearby surfaces and equipment. Protect structures located close to a brazing site with a heat shield.

Another fire hazard exists when brazing connections close on a line that pass through an insulated wall. Some of the pressed insulation sheets can begin to smoulder when accidentally brushed by the hot flame of a torch. The insulation may begin to smoulder unnoticeably until the heat gradually builds-up in a wall and causes spontaneous combustion. Protect the entry and exit openings of an insulated wall with damp rags when brazing nearby connections.

Occasionally a leak may develop in a brazing outfit where the regulator is connected to the fuel gas tank, at the packing gland nut of valve stem or where the hose connects to the torch or regulator. Each fuel gas is formulated to have its own characteristic odor which facilitates leak detection. Each is easily recognizable. For instance, acetylene has a distinctive garlic-like odor. Other fuel gases have an ether-like odor or smell like the natural gas used in home cooking and heating. A keen nose can be invaluable in heading off a potential disaster. In the event your olfactory senses indicate the presence of fuel gas, use refrigerant leak detection solution to pinpoint the leak. In most cases a leak can be fixed by merely tightening a connection to effect an air tight seal.

GLOSSARY

ACR COPPER TUBE—Copper tube made specifically for the air conditioning and refrigeration trade. Special attention is given to internal cleanliness.

AWS—Abbreviation for American Welding Society.

BLANCHING—The characteristic color changes seen when heating copper tubing above 600°F in the open atmosphere with a reducing flame.

BRAZING—A method of joining metal by means of a brazing alloy which melts above 840°+ but below the melting point of the base metals (i.e., the parts to be joined). The brazing alloy wets the surface layers of the metallic parts to form an alloy bond.

BRAZING ALLOY—An alloy used in brazing which has a melting temperature above 840°F.

BRAZE SHAPING—The ability to move and shape a partially molten brazing alloy into position in a brazed joint.

BRAZE SPREADING—The ability of a brazing alloy to spread in all directions and wet the base metal when molten.

BRAZE STRENGTHENING—The ability to use the inherent strength of a brazing alloy to build up and reinforce weak sections of a brazed joint.

CAPILLARY ACTION—The force that results from adhesion of a liquid to a solid surface which is greater than the internal cohesion of the liquid itself.

CAPILLARY BRAZING—A brazing method that relies primarily upon the force of capillary action to draw molten brazing alloy into the clearance space between the mating metallic surfaces of coupling or joint.

CAPILLARY SPACE OR GAP—The space between the outside surface of a tube-end inserted into a sweat joint and the internal surface of the joint. This gap promotes capillary flow which draws molten filler metal into the far recesses of a joint. (Also known as clearance space.)

CLEARANCE VOLUME—The total volume between the cup and spigot which needs to be filled with filler metal.

CUP or BELL—Typically the enlarged opening of a fitting (female end) which receives the tube-end (male end) in a sweat joint.

EUTECTIC ALLOY—An alloy with a sharply defined melting point.

FILLER METAL—A metal or alloy employed in soldering, brazing or welding to join metal surfaces.

FILLET BRAZING—A brazing practice wherein only a fillet seals the brazing mouth of a joint to bond the components of the connection together.

FLOW POINT—The temperature at which a brazing alloy is sufficiently molten to flow and fill the capillary space of a joint.

FLUX—A chemical used to promote wetting between a filler metal and its base metal.

LIQUATION—This condition is the fractional separation of a brazing alloy, which exhibits a wide melting range, caused by slowly heating the alloy. The slow heating process allows the liquid phases, with lower melting points, to sequentially drain away from the main body of the alloy. Once the separation occurs, only residues with higher melting points are left behind.

LIQUIDUS—Latin term that designates the lowest temperature at which an alloy remains completely liquid. Full or partial solidification takes place below the liquidus.

MELTING POINT—The temperature at which a brazing alloy begins to melt.

NEUTRAL FLAME—A flame which results from complete combustion of fuel and oxygen.

OXIDIZING FLAME—An oxygen-fuel gas flame adjusted to contain an excess of oxygen (i.e., lean mixture).

REDUCING FLAME—An oxygen-fuel gas flame adjusted to contain an excess of fuel gas (i.e., rich mixture).

SLOT OR MOUTH—The annular opening between the tube- and cup-end into which the molten filler metal is fed.

SOLDER OR BRAZING RING—An annular ring of filler metal which appears around the solder or brazing slot after the clearance volume has been completely filled.

SOLDERING—Soldering is a metal joining process which uses a soldering alloy that has a melting temperature below 840°F and below that of the parts to be joined. Molten solder alloy wets the surface layers to form an alloy bond as it is drawn into the clearance space by capillary action.

SOLIDUS—Latin designation for the highest temperature at which an alloy remains completely solid. Full or partial melting (i.e., pasty stage) takes place above the solidus.

SPIGOT—The tube-end which is inserted into the fitting cup.

SWEAT—The generic term applied when describing piping connections which are brazed or soldered.

TARGET HEAT—The portion of the heat in a torch flame that can be directed on the work to effect the brazing or soldering operation.

SWIRL TIP—A torch tip that operates on the swirl combustion principle. The fuel-air mixture passes through a series of deflection plates which imparts a rotary mixing motion to the mixture. This turbulence causes the mixture to burn at a hotter temperature and results in a condensed flame.

WETTING—The process wherein a permanent alloy bond is formed between a base metal and its filler metal.

APPENDIX

Table 1. ACR tubing used in the air conditioning and refrigeration trades.

Size, inches	Nominal Dimensions, inches			Calculated Values, Based on Nominal Dimensions			
	Outside Diameter	Inside Diameter	Wall Thickness	Cross Sectional Area of Bore, sq. inches	External Surface, sq. ft. per lin. ft.	Internal Surface, sq. ft. per lin. ft.	pounds per lin. ft.
1/8	**.125**	**.065**	**.030**	**.00332**	**.0327**	**.0170**	**.0347**
3/16	**.188**	**.128**	**.030**	**.0129**	**.0492**	**.0335**	**.0577**
1/4	**.250**	**.190**	**.030**	**.0284**	**.0655**	**.0497**	**.0804**
5/16	**.312**	**.248**	**.032**	**.0483**	**.0817**	**.0649**	**.109**
3/8	.375	.315	.030	.0780	.0982	.0821	.126
3/8	**.375**	**.311**	**.032**	**.0760**	**.0982**	**.0814**	**.134**
1/2	**.500**	**.436**	**.032**	**.149**	**.131**	**.114**	**.182**
1/2	.500	.430	.035	.145	.131	.113	.198
5/8	**.625**	**.555**	**.035**	**.242**	**.164**	**.145**	**.251**
5/8	.625	.545	.040	.233	.164	.143	.285
3/4	.750	.666	.042	.348	.196	.174	.362
7/8	.875	.785	.045	.484	.229	.206	.455
1 1/8	1.125	1.025	.050	.825	.294	.268	.655
1 3/8	1.375	1.265	.055	1.26	.360	.331	.884
1 5/8	1.625	1.505	.060	1.78	.425	.394	1.14
2 1/8	2.125	1.985	.070	3.09	.556	.520	1.75
2 5/8	2.625	2.465	.080	4.77	.687	.645	2.48
3 1/8	3.125	2.945	.090	6.81	.818	.771	3.33
3 5/8	3.625	3.425	.100	9.21	.949	.897	4.29
4 1/8	4.125	3.905	.110	12.0	1.08	1.02	5.38

NOTE: Sizes shown in bold face type are available in annealed temper only; sizes shown in italics are available in hard temper only, all others in both.

Table 2. Allowable stress for annealed copper tube at indicated temperatures (Courtesy, Mueller Brass Co.).

Table gives computed allowable stress for annealed copper tube at indicated temperatures
SAFE WORKING INTERNAL PRESSURES — ANSI-B31.5-1974

Tube OD	Wall Thickness	Wt. Per Foot	Lengths Per Bundle	150°F PSI	250°F PSI	350°F PSI	400°F PSI
3/8	.030	.126	100	760	700	600	460
1/2	.035	.198	25	680	620	520	400
5/8	.040	.285	25	630	580	490	370
3/4	.042	.362	10	550	510	430	320
7/8	.045	.455	10	500	460	390	300
1 1/8	.050	.655	5	430	400	340	250
1 3/8	.055	.884	5	390	360	300	230
1 5/8	.060	1.14	5	370	340	280	220
2 1/8	.070	1.75	3	310	290	240	180
2 5/8	.080	2.48	2	300	280	230	170
3 1/8	.090	3.33	1	280	260	220	170
3 5/8	.100	4.29	1	270	250	210	160
4 1/8	.110	5.38	1	250	230	200	140
5 1/8	.123	7.61	1	240	220	190	140
6 1/8	.140	10.20	1	220	200	170	130

Table 3. Common portable fuel tank sizes (Courtesy, L-TEC Welding & Cutting Systems).

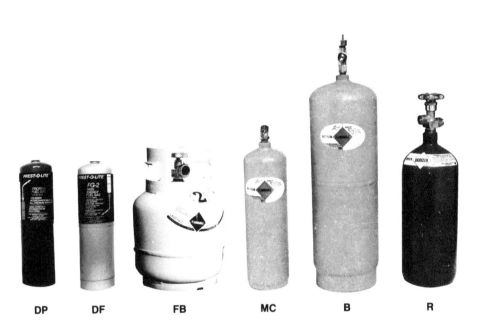

DP   DF   FB   MC   B   R

SPECIFICATIONS

Style	Capacity	Height	Diameter	Weight, Full
Disposable Propane (DP)	14.1 oz.	10½ in.	2¾ in.	2 lb., 2 oz.
Disposable FG-2 (DF)	15.2 oz.	10½ in.	2¾ in.	2 lb., 3 oz.
Refillable FG-2 (FB)	7½ lb.	13¼ in.	9⅛ in.	10¾ lb.
Type MC Acetylene	10 ft^3	14 in.	4 in.	8 lb.
Type B Acetylene	40 ft^3	23 in.	6¼ in.	26 lb.
Type R* Oxygen	20 ft^3	19 in.	5³⁄₁₆ in.	13½ lb.

*Also available for nitrogen service

Note 1: Analysis shall be made for the elements for which specific values are shown in this table. If, however, the presence of other elements is indicated in the course of routine analysis, further analysis shall be made to determine that the total of these other elements is not present in excess of the limits specified for "other elements total" in the last column in the table.

Note 2: Single values shown are maximum percentages, except where otherwise specified.

Table 4. American Welding Society (AWS) specifications for brazing filler metals (Used with permission from *AWS Brazing Manual*).

AWS classification	Silver								
	Ag	Cu	Zn	Cd	Ni	Sn	Li	P	Other elements total
BAg-1	44.0-46.0	14.0-16.0	14.0-18.0	23.0-25.0	—	—	—	—	0.15
BAg-1a	49.0-51.0	14.5-16.5	14.5-18.5	17.0-19.0	—	—	—	—	0.15
BAg-2	34.0-36.0	25.0-27.0	19.0-23.0	17.0-19.0	—	—	—	—	0.15
BAg-2a	29.0-31.0	26.0-28.0	21.0-25.0	19.0-21.0	—	—	—	—	0.15
BAg-3	49.0-51.0	14.5-16.5	13.5-17.5	15.0-17.0	2.5-3.5	—	—	—	0.15
BAg-4	39.0-41.0	29.0-31.0	26.0-30.0	—	1.5-2.5	—	—	—	0.15
BAg-5	44.0-46.0	29.0-31.0	23.0-27.0	—	—	—	—	—	0.15
BAg-6	49.0-51.0	33.0-35.0	14.0-18.0	—	—	—	—	—	0.15
BAg-7	55.0-57.0	21.0-23.0	15.0-19.0	—	—	4.5- 5.5	—	—	0.15
BAg-8	71.0-73.0	remainder	—	—	—	—	—	—	0.15
BAg-8a	71.0-73.0	remainder	—	—	—	—	0.25-0.50	—	0.15
BAg-13	53.0-55.0	remainder	4.0- 6.0	—	0.5-1.5	—	—	—	0.15
BAg-13a	55.0-57.0	remainder	—	—	1.5-2.5	—	—	—	0.15
BAg-18	59.0-61.0	remainder	—	—	—	9.5-10.5	—	—	0.15
BAg-19	92.0-93.0	remainder	—	—	—	—	0.15-0.30	0.025	0.15
BAg-20	29.0-31.0	37.0-39.0	30.0-34.0	—	—	—	—	—	0.15
BAg-21	62.0-64.0	27.5-29.5	—	—	2.0-3.0	5.0- 7.0	—	—	0.15

Table 4. (Continued)

| AWS classi-fication | 4B Copper-Phosphorus[a] | | | |
	P	Ag	Cu	Other elements total
BCuP-1	4.8-5.2	—	remainder	0.15
BCuP-2	7.0-7.5	—	remainder	0.15
BCuP-3	5.8-6.2	4.8- 5.2	remainder	0.15
BCuP-4	7.0-7.5	5.8- 6.2	remainder	0.15
BCuP-5	4.8-5.2	14.5-15.5	remainder	0.15
BCuP-6	6.8-7.2	1.8- 2.2	remainder	0.15
BCuP-7	6.5-7.0	4.8- 5.2	remainder	0.15

[a]Chemical analysis for phosphorus contents of these ranges should be made by any suitable method agreed upon by the supplier and purchaser. In case of dispute, the procedure in the latest edition of the ASTM E156, Standard Photometric Method for Determination of Phosphorus in High-Phosphorus Brazing Alloys shall be the referee method.

Table 5. Common brazing alloys (Courtesy, Engelhard).

AWS A5.8	Engelhard	J. W. Harris	Handy & Harman	Nominal Chemical Composition							Melting Point °F	Flow Point °F	Recommended for Basic Metals*
				Ag	Cu	Zn	Cd	P	Ni	Sn			
BCuP-2	Silvaloy 0	Stay-Silv 0 **(0)	Flos-Flo 7	0	92.75			7.25			1310	1350	1
—	Silvaloy 1	— (1)	—	1	93			6			1190	1350	1
BCuP-6	Silvaloy 2	Stay-Silv 2 (2)	Sil-Fos 2	2	91			7			1190	1350	1
BCuP-3	Silvaloy 5	Stay-Silv 5 (5)	Sil-Fos 5	5	89			6			1190	1350	1
—	Silvaloy 6	Stay-Silv 6 (6)	Sil-Fos 6	6	87.75			7.25			1190	1350	1
BCuP-5	Silvaloy 15	Stay-Silv 15 (15)	Sil-Fos 15	15	80			5			1190	1300	1
BAg-1a	Silvaloy 50	Not Available (50)	Easy-Flo 50	50	15.5	16.5	18				1160	1175	1, 2, 3, 4
BAg-1	Silvaloy 45	Not Available (45)	Easy-Flo 45	45	15	16	24				1125	1145	1, 2, 3, 4
BAg-2	Silvaloy 35	Not Available (35)	Easy-Flo 35	35	26	21	18				1125	1295	1, 2, 3, 4
BAg-2a	Silvaloy 30	Not Available (30)	Easy-Flo 30	30	27	23	20				1125	1310	1, 2, 3, 4
BAg-3	Silvaloy 50N	Not Available (503)	Easy-Flo 3	50	15.5	15.5	16		3		1170	1270	1, 2, 3, 4
BAg-6	Silvaloy A-50	Safety-Silv 1425 (A-25)	Braze 501	50	34	16					1250	1425	1, 2, 3
BAg-5	Silvaloy A-45	Safety-Silv 1370 (A-18)	Braze 450	45	30	25					1225	1370	1, 2, 3
—	Silvaloy A-40L	Safety-Silv 1350 (A-295)	Braze 401	40	30	30					1245	1340	1, 2, 3
—	Silvaloy A-35	Safety-Silv 1350G (A-22)	—	35	32	33					1150	1350	1, 2, 3
BAg-20	Silvaloy A-30	Safety-Silv 1410 (A-13)	Braze 300	30	38	32					1250	1410	1, 2, 3
BAg-13	Silvaloy A-54N	Safety-Silv 1575 (54)	Braze 541	54	40	5			1		1325	1575	1, 2, 3, 4
BAg-24	Silvaloy A-50N	Safety-Silv 1305 (502)	Braze 505	50	20	28			2		1220	1305	1, 2, 3, 4
BAg-4	Silvaloy A-40N2	Safety-Silv 1435 (250)	Braze 403	40	30	28			2		1220	1435	1, 2, 3, 4
—	Silvaloy A-40N5	Safety-Silv 1580 (254)	Braze 404	40	30	25			5		1220	1580	1, 2, 3, 4
BAg-7	Silvaloy A-56T	Safety-Silv 1200 (355)	Braze 560	56	22	17				5	1145	1205	1, 2, 3, 4
BAg-28	Silvaloy A-40T	— (40T)	Braze 402	40	30	28				2	1200	1310	1, 2, 3, 4
—	Silvaloy A-25T	Safety-Silv 1375 (25T)	Braze 255	25	40	33				2	1265	1400	1, 2, 3

*1 — Copper and Copper Alloys
2 — Nickel and Nickel Alloys
3 — Carbon Steel
4 — Stainless Steel
**Old names appear in parenthesis.

Table 6. Copper-phosphorus brazing alloys manufactured by Engelhard (Courtesy, Engelhard).

SILVALOY	% Ag	% P	% Cu	°F Solidus	°F Liquidus	°F Flow Point	AWS Classification
15	14.5-15.5	4.8 -5.2	Balance	1190	1475	1300	BCuP-5
6	5.8- 6.2	5.75-6.25	Balance	1190	1485	1325	–
6F	5.8- 6.2	7.0 -7.5	Balance	1190	1325	1350	BCuP-4
5	4.8- 5.2	5.8 -6.2	Balance	1190	1490	1325	BCuP-3
5F	4.8- 5.2	6.5 -7.0	Balance	1190	1420	1300	BCuP-7
2M	1.8- 2.2	6.25-6.75	Balance	1190	1460	1350	–
2	1.8- 2.2	6.8 -7.2	Balance	1190	1450	1300	BCuP-6
1	0.8- 1.2	5.8 -6.2	Balance	1190	1495	1325	–
0M	–	6.55-7.05	Balance	1310	1470	1350	–
0	–	7.0 -7.5	Balance	1310	1460	1350	BCuP-2

NOTE: The naming system's number indicates the nominal percentage of silver in the alloy. The letters M or F indicate flow characteristics. For instance Silvaloy 6F contains 6% silver but has greater flow characteristics than Silvaloy 6. The letter M indicates less flow characteristics.

Table 7. Brazing alloys manufactured by the J. W. Harris Company (Courtesy, J. W. Harris Co.)

ALLOY	SILVER %	PHOS. %	MELTING RANGE SOLIDUS	MELTING RANGE LIQUIDUS	FLUIDITY RATING*	RECOMMENDED JOINT CLEARANCE	SPECIFICATION** AWS A5.8-81	SPECIFICATION** FED QOB650B	.031 1/32	.047 3/64	.062 1/16	.093 3/32
Stay-Silv 0	0	7.10	1310	1475	5	.002/.005	$BCuP_2$	$BCuP_2$	4275	1900	1070	475
Stay-Silv 0LP	0	6.80	1310	1510	4	.003/.005			4275	1900	1070	475
Stay-Silv 0HP	0	7.40	1310	1445	6	.002/.005	$BCuP_2$	$BCuP_2$	4275	1900	1070	475
Brayzon	1	7.00	1190	1465	4	.003/.005			4275	1900	1070	475
Stay-Silv 2	2	7.00	1190	1450	4	.003/.005	$BCuP_6$		4275	1900	1070	475
Stay-Silv 2LP	2	6.60	1190	1500	3	.003/.006			4275	1900	1070	475
Stay-Silv 2HP	2	7.40	1190	1405	5	.002/.005			3425	1900	1070	475
Stay-Silv 5	5	6.00	1190	1500	3	.003/.006	$BCuP_3$		4275	1900	1070	475
Stay-Silv 5LP	5	5.70	1190	1535	2	.004/.008		$BCuP_3$	4275	1900	1070	475
Stay-Silv 5HP	5	6.50	1190	1445	4	.003/.005	$BCuP_7$		4275	1900	1070	475
Stay-Silv 6	6	6.50	1190	1425	5	.002/.005			4275	1900	1070	475
Stay-Silv 6LP	6	6.20	1190	1455	4	.003/.005			4275	1900	1070	475
Stay-Silv 6HP	6	7.20	1190	1335	7	.001/.005	$BCuP_4$	$BCuP_4$	4275	1900	1070	475
Dynaflow	6	6.10	1190	1465	3	.003/.006			4275	1900	1070	475
Stay-Silv 15	15	5.00	1190	1480	3	.003/.006	$BCuP_5$	$BCuP_5$	4160	1850	1040	460
Stay-Silv 15LP	15	4.70	1190	1515	2	.004/.008			4160	1850	1040	460
Stay-Silv 15HP	15	5.40	1190	1435	4	.003/.005			4160	1850	1040	460
Phoson +	15	7.30	1190	1205	10	.001/.003			4160	1850	1040	460

*High numbers indicate fluidity within the melting range.
**Stay-Silv 15 also meets Fed Spec QOB654A, Grade III and Military Spec B-15395-A, Grade III.

Table 8. Estimated quantities of brazing alloy required per joint (Courtesy, J. W. Harris Co.)

ESTIMATED AMOUNTS OF BRAZING ALLOY REQUIRED PER JOINT:

Normal Tubing Size	3/64" Wire	1/16" Wire	3/32" Wire	1/8 x .050 Rod	Tip Size No.	Estimated Acetylene Use (C.F.H.)
1/4"	1 1/4"	3/4"			4	10- 17
3/8"	1 1/2"	1"			4	10- 17
1/2"	2"	1 1/2"	3/4"	7/8"	5	17- 30
3/4"	3"	2"	1"	1 1/8"	5	17- 30
1"		3"	1 1/2"	1 5/8"	6	30- 40
1 1/4"		4"	2"	2 1/2"	6	30- 40
1 1/2"			2 1/2"	2 3/4"	7	40- 50
2"			3 3/4"	4 1/2"	8	50- 75
2 1/2"			6"	7 1/2"	8	50- 75
3"			10"	11 1/2"	9	65- 90
3 1/2"			12"	13 3/4"	9	65- 90
4"			14"	16"	10	75-100
6"			21"	23 3/4"	10	75-100
A	1900	1068	475		513 in. of alloy/lb.	
B	118	67	29		inches of alloy per troy oz.	

A— Phos copper silver alloys. Dynaflow®, Stay-Silv 15
B— Silver Brazing alloys. Stay-Silv 45, Safety Silv alloys.
The above figures are approximate and will vary depending on joint clearance, depth and operator technique.

Table 9. Copper-phosphorus brazing alloys manufactured by Handy & Harmon (Courtesy, Handy & Harman).

Filler Metal Name	Nominal Composition %				Description and Application (See note below)
	Ag	P	Cu	Total other elements, max.	
Sil-Fos 18	18	7.25	Bal	0.15	A ternary eutectic filler metal for joints where good fit-up can be maintained and low melting point is of prime importance. Clearance: .001" to .003."
Sil-Fos	15	5	Bal	0.15	For use where close fit-ups cannot be maintained and joint ductility is important. Recommended joint clearance: .001" to .005."
Sil-Fos 6	6	7.25	Bal	0.15	A very fluid filler metal for close fit-up work. Low melting range makes it ideal where temperature is a factor. Recommended joint clearance: .001" to .003."
Sil-Fos 6M	6	6	Bal	0.15	Recommended for use where close fit-up cannot be maintained. Has the ability to fill gaps and form fillets without affecting joint strength. Recommended joint clearance: .002" to .005."
Sil-Fos 5	5	6	Bal	0.15	Designed primarily for those applications where close fit-ups cannot be maintained. It has ability to fill gaps and form fillets without adversely affecting joint strength. Recommended joint clearance: .003" to .005."
Sil-Fos 5F	5	6.75	Bal	0.15	Designed to fill the gap between Sil-Fos 5 and Sil-Fos 6. It is more fluid than Sil-Fos 5, but not as fluid as Sil-Fos 6. It has the ability to fill small gaps without adversely affecting joint strength. Recommended joint clearance: .002" to .005."
Sil-Fos 2	2	7	Bal	0.15	A filler metal with comparable characteristics to Fos-Flo 7 when heated rapidly. With slower heating it becomes similar to Sil-Fos and Sil-Fos 5. Recommended joint clearance: .001" to .005."
Sil-Fos 2M	2	6.6	Bal	0.15	Has ability to fill moderate gaps in poorly fitted joints. More ductile than Fos-Flo 7 or Sil-Fos 2. Intended for use on copper tube headers and similar applications where a sleeve fit is not practical. Recommended joint clearance: .001" to .005."
Sil-Fos 1	0.9	6.5	Bal	0.15	Low silver filler metal similar to Sil-Fos 2M, but with slightly higher brazing temperature. Recommended joint clearance: .001" to .005".
Fos-Flo 7	—	7.25	Bal	0.15	An economical, very fluid medium temperature filler metal for use with copper, brass and bronze. Withstands moderate vibration. Recommended joint clearance: .001" to .003."
Fos-Flo 5	—	5	Bal	0.15	A low cost filler metal with wide melting range and low fluidity. Used primarily for preplacing in connection with resistance welding or to supplement spot welding operations. Recommended joint clearance: .003" to .005."

NOTE: For use with copper and copper alloy base metals. Do not use to join ferrous materials as brittle phosphide compounds will be formed.

Table 9. (Continued)

Solidus	Liquidus	Brazing Range	Flow Characteristics	ASME Blr. & Pr. Vsl. Cd., Sec. II-C, SFA5.8, 1983 and AWS A5.8	Fed. Spec. QQ-B-650B	QQ-B-654A Amend. 1
				Specifications		
1190°F (645°C)	1190°F (645°C)	1200°F to 1300°F (650°C) (705°C)	Very fast			
1190°F (645°C)	1475°F (800°C)	1300°F to 1500°F (705°C) (815°C)	Slow	BCuP-5	BCuP-5	BCuP-5 or Grade III*
1190°F (645°C)	1325°F (720°C)	1275°F to 1450°F (690°C) (790°C)	Fast	BCuP-4	BCuP-4	
1190°F (645°C)	1460°F (795°C)	1300°F to 1500°F (705°C) (815°C)	Slow			
1190°F (645°C)	1495°F (815°C)	1325°F to 1500°F (720°C) (815°C)	Slow	BCuP-3	BCuP-3	
1190°F (645°C)	1420°F (770°C)	1300°F to 1500°F (705°C) (815°C)	Medium	BCuP-7		
1190°F (645°C)	1450°F (785°C)	1325°F to 1500°F (720°C) (815°C)	Medium	BCuP-6		
1190°F (645°C)	1495°F (815°C)	1350°F to 1550°F (730°C) (845°C)	Slow			
1190°F (645°C)	1500°F (815°C)	1350°F to 1600°F (730°C) (870°C)	Slow			
1310°F (710°C)	1460°F (795°C)	1350°F to 1550°F (730°C) (845°C)	Fast	BCuP-2	BCuP-2	
1310°F (710°C)	1695°F (925°C)	1450°F to 1700°F (790°C) (925°C)	Slow	BCuP-1	BCuP-1	

Table 10. Typical manufacturer's tip selection guides for copper tube fittings.

Tip Size			Gas Flow		Copper Tubing Size Capacity Propane Gas				Copper Tubing Size Capacity MAPP Gas			
			@ 24 psi	@ 36 psi	Soft Solder		Silver Solder		Soft Solder		Silver Solder	
Tip No.	in.	mm.	lbs./hr.	lbs./hr.	in.	mm.	in.	mm.	in.	mm.	in.	mm.
T-2	5/16	8.0	.14	.17	1/8-3/8	3-10	1/16-1/2	1-6	1/8-1/2	3-12	1/8-3/8	3-10
T-3	7/16	11.1	.20	.25	1/4-1	6-25	1/8-3/8	3-10	3/8-1 1/2	10-40	1/4-5/8	6-15
T-4	1/2	12.7	.39	.48	3/8-1 1/2	10-40	1/4-5/8	6-15	3/8-2 1/2	10-60	1/4-1 3/8	6-35
T-5	3/4	19.0	1.10	1.3	1 1/2-2 1/2	25-60	5/8-1 3/8	15-35	1-4	25-100	5/8-2 1/8	15-50
T-6	1	25.4	2.10	2.5	1 1/2-4	40-100	7/8-2 1/8	20-50	1 1/2-6	25-150	7/8-4	20-100

"Total LP" torch tips (Courtesy, Wingaersheek).

Tip No.	Flame Dia.	Type Flame	Soft Solder (Inches)	Silver Solder (Inches)
BA-1	1/16	Pencil	1/6-1/8	1/6-1/8
BA-2	1/8	Pencil	1/8-1/4	1/8-1/4
BA-3	3/16	Pencil	1/4-1/2	1/4-1/2
BA-4	1/4	Brush	1/2-1	1/2-3/4
BA-5	5/16	Brush	1-1 1/4	3/4-1
BA-6	7/16	Brush	1 1/4-1 1/2	1-1 1/4
BA-10	3/8	Swirl	1-3	Up to 2
BA-12	1/2	Swirl	1-5	Up to 3

Acet-O-Lite tips (Courtesy, Goss, Inc.)

Table 10. (Continued)

Fitting Size, in.	Soft-Solder			Silver-Braze		
	Standard Tip	SWIRLJET Acetylene	SWIRLJET Propane	Standard Tip	SWIRLJET Acetylene	SWIRLJET Propane
1/2	3	3	3	5	3	4
3/4	4	3	4	5	3	4
1	5	3	4	6	3	5
1 1/2	6	4	4	6	4	5
2	6	4	5	—	5	5
3	6	5	5	—	5	—
4	6	5	5	—	5	—

Prest-O-Lite Tips (Courtesy, L-TEC Welding & Cutting Systems).

INDEX

Make The Right Connection

Need a solid working background on valves and their proper applications—this is the source book you've been waiting for!

Valve Selection & Service Guide

Hard cover, 202 pages

This guide was written for contractors, servicemen and anyone who must size, select and install or repair valves on the job. The information in this book targets the air conditioning, heating and refrigeration industry, but the general principles and basic valve designs covered apply to a number of other fields as well.

Many manufacturers can help you select the right valve design required for your job, but the recommendations they make are based on the information you provide. If your information is sketchy or not thoroughly planned out, improper selections are likely to occur. The result, equipment failure, or even worse, personal injury and/or a lawsuit that can translate into considerable financial loss.

Get the Valve Selection and Service Guide so you can learn to select the proper valves for your applications and **prevent costly mistakes.** It may prove to be the best investment you'll ever make!

Make the right connection...

The Schematic Wiring Book Set!

Remember connect the dots?
Then try your hand at connecting this circuit so
the lamp lights when either switch is closed.

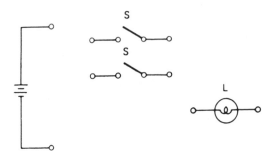

If this simple task has you stumped, you need
the *Schematic Wiring Book Set*. Because once
you learn electrical shorthand, you'll be able to
read any manufacturer's wiring diagram.

Troubleshoot electrical systems like a pro!